A Century of Physics

Springer

New York
Berlin
Heidelberg
Barcelona
Hong Kong
London
Milan
Paris
Singapore
Tokyo

A Century of Physics

D. Allan Bromley

Sterling Professor of the Sciences and Dean of Engineering
Yale University
New Haven, Connecticut

 Springer

D. Allan Bromley
Yale University
P.O. Box 208124
New Haven, CT 06520
USA

Library of Congress Cataloging-in-Publication Data
Bromley, D. Allan (David Allan), 1926–
    A century of physics / D. Allan Bromley.
        p. cm.
    ISBN 0-387-95247-0 (alk. paper)
    1. Physics—History—20th century.   I. Title.
QC7 .B68 2001
530′.09′04—dc21
                            2001018410

Printed on acid-free paper.

A substantially abbreviated version of this book was presented at the opening of the American Physical Society Centennial Meeting in Atlanta, Georgia, on March 21, 1999.

Production managed by A. Orrantia; manufacturing supervised by Jacqui Ashri.
Photocomposed pages prepared by Matrix Publishing Services, Inc., York, PA.
Printed and bound by Walsworth Publishing Co., Brookfield, MO.
Printed in the United States of America.

9 8 7 6 5 4 3 2 1

ISBN 0-387-95247-0   SPIN 10794156

Springer-Verlag   New York   Berlin   Heidelberg
A member of BertelsmannSpringer Science+Business Media GmbH

To Vickie

Preface

When I was invited to deliver the opening plenary talk at the Centennial Meeting of the American Physical Society, scheduled for Atlanta in March 1999, the meeting was two years in the future and in classic ancient Samoan fashion (counting one, two, three, infinity) I felt that I had an almost infinite time to prepare. How short those two years actually were!

In September of 1998 I wrote to a large number of my friends asking for their input and assistance and both were given generously. So also were historically significant photographs, and I am especially grateful to Spencer Weart of the AIP Center for the History of Physics for giving me access to its Emilio Segrè Visual Archives. Fortunately, too, I have been collecting such photographs myself over the years and so I had on hand many of those included here.

Particularly in view of the availability of these photographs, I have chosen to treat the first half of our century in physics from an historical point of view, focusing on the major discoveries that provided the foundation for the explosive growth of physics—and indeed all of science—in the second half of our century.

The fragmentation that accompanied this growth is reflected in my somewhat arbitrary selection of specific areas of discovery, and application of physics, in the post-World War II period. The selection was necessarily a very personal one, and I recognize that I have not given many areas of physics—for example, optics and acoustics—the emphasis that they deserve.

Happily, as we come to the end of our century, it has become clear—as I hope will become clear to the reader—that after this period of increasing fragmentation, during which many of my colleagues have despaired about our future in physics, the fundamental underlying unity of our science has begun to reassert itself in area after area.

Preparing this report has been a very educational process, and my only regret is that space and time limitations prevented me from including much excellent physics that I would have liked to bring to my reader's attention.

In a global review of a major scientific field such as physics, I have found that to include specific references was simply not possible; instead, I have attempted to include names and institutions in the text so that follow-up can be accomplished when and if desired. If I were to list, in a report such as this, the names of all those whose work I have quoted and from whom I have received information and illustrations and with whom I have had useful discussions, the list would occupy more space than the report itself. As at the outset, however, I would take this opportunity to thank all of those who have been so generous in providing assistance, and my sincere apologies go to all those whose work I have not been able to include.

Finally, let me thank Michael Skrzyniarz and Mary Anne Schulz for all their help in assembling this report. It would not have been possible without them. And it has been a pleasure working with Tom von Foerster of Springer Verlag.

D. Allan Bromley
New Haven, Connecticut
May, 1999

Contents

PART I
An Historical Overview, 1900–1949

Introduction

As suggested by Figure 1, science and its applications—which we would today call technology—have been valued in American society from the very beginning. As we approach the close of the twentieth century, it is entirely appropriate that we celebrate the role of our particular sector of this science and technology—that is, physics and its applications. It is obviously impossible to cover so rich a field with anything approaching completeness, so I will apologize in advance to all those who feel that their work has been slighted. It has often been said that one picture is worth a thousand words, I shall therefore include a large number of pictures. My goal in this part of this report is to remind the reader of some of the high points of physics during the first half of our century that laid the foundation for modern physics and of the importance of these discoveries to society and to the human race in general.

What Is Physics?

But what, then, is physics? Probably the best definition that I have encountered is that shown in Figure 2, written in 1970 by an old (and unfortunately late) friend, Ed Purcell. "Science is knowing." Purcell said. What man knows about inanimate nature is physics, or rather the most lasting and universal things that he knows make up physics." This is a very important part of a definition of physics. We physicists have the arrogance to believe that the laws we deduce from our measurements here on earth apply throughout the universe and that what is true today was true throughout the entire life of that universe; and our measurements support that arrogance. Purcell went on to say that "As [man] gains more knowledge what would have appeared complicated or capricious can be seen as essentially simple and in a deep sense orderly." I know that Pur-

"—There is nothing which can better deserve your
patronage than the promotion of Science and Literature.
Knowledge is in every country the surest basis of public
happiness."

George Washington
State of the Union Address
January 8, 1790

FIGURE 1

cell considered adding: "and in the deepest sense, beautiful." And he would
have also been entirely correct in saying that.

Turning to applications, Purcell said, "to understand how things work is to
see how, within environmental constraints and the limitations of wisdom, better to accommodate nature to man and man to nature." This brief statement
says, or at least *should* say, much about modern technology.

The Early Twentieth Century

Many in the past have noted that, for science, the twentieth century really began in 1897 with J.J. Thompson's discovery of the electron (see Figure 3), a
judgment that reflects the enormous impact our ability to manipulate the atom
and its component electrons has had on areas including communications, computation, energy, medicine, and almost every area of modern civilization.

This twentieth century in physics began with a rush of new insights, and,

"Science is knowing. What man knows about inanimate nature
is physics, or rather the most lasting and universal things that he
knows make up physics"

"… that has enabled man to understand more of what goes on in
his universe. As he gains more knowledge what would have
appeared complicated or capricious can be seen as essentially
simple and in a deep sense orderly"

"… to understand how things work is to see how, within
environmental constraints and the limitations of wisdom, better
to accommodate nature to man and man to nature"

Edward M. Purcell
Physics in Perspective
National Academy of Science, 1970

FIGURE 2

FIGURE 3   *J.J. Thomson (1906)* at work in his laboratory in the 1890s with the appara-tus that he used to determine the ratio of the electron's electrical charge to its mass. (Copyright University of Cambridge, Cavendish Laboratory.)*

happily, it is ending in much the same way—as we shall see. In 1905, for ex-ample, Albert Einstein published his classic papers on Brownian motion, on the photoelectric effect, and on special relativity, and provided one of the classic equations of all time in $E = mc^2$, his statement that mass is frozen energy. It was not until 1915 that he published his general theory of relativity, a publica-tion that established his reputation as the outstanding figure in physics in the twentieth century. Figure 4 shows him with four other Nobel laureates, and Fig-ure 5 shows him together with his old friend, and friendly adversary, Niels Bohr. These two argued throughout their lifetimes about the meaning of quantum mechanics and never reached agreement.

Activities in the Cambridge Cavendish Laboratory

In 1911, Ernest Rutherford discovered the atomic nucleus. Rutherford was not a particularly good experimentalist, but he had an uncanny ability to think of simple experiments that were capable of providing profound results. Figure 6 shows him in his laboratory with his lifelong technical assistant, Ratcliffe. The

---

*Years in parentheses indicate when the individual received The Nobel Prize.

FIGURE 4   *Nobel Laureates, 1928. From the left: Walter Nernst (1920), Albert Einstein (1921), Max Planck (1918), Robert Milliken (1923), and Max von Laue (1914). Courtesy of the AIP Emilio Segre Visual Archives.)*

sign above them is a reminder that the instrumentation of the time was so microphonic that almost any sound could destroy the measurement in progress.

The physics community, at least within Europe, was a very flexible and mobile one, where individuals moved to whatever laboratory, university, or group they felt to be doing the most interesting work at any given moment. This mobility led directly to some of the most important developments in all of science and especially in physics.

As a particular example, Niels Bohr visited Ernest Rutherford in Cambridge (see Figure 7) and in 1913 published his description of what has come to be known as the "Bohr-Rutherford" model of the atom, wherein electrons move in circular orbits about the Rutherford nucleus. This model was remarkably successful in reproducing the pattern of the spectral lines of hydrogen, particularly when the model was fine-tuned by Sommerfeld, who posited electrons moving in elliptical rather than circular orbits (see Figure 8). Both models, of course, stemmed from the seminal ideas that had been published by Max Planck (see Figure 9), who had first introduced the concept of the quantum of energy in his efforts to understand the spectrum of the radiation from a black body.

Substantial progress toward understanding the further complexity of atomic spectra resulted from the work of Uhlenbeck and Goudsmidt (see Figure 10), who first introduced the concept of electron spin. In Figure 10 they are shown with Henrik Kramers—distinguished not only by his physics, but by his bowtie!

FIGURE 5   *Albert Einstein (1921) and Niels Bohr (1922) in Brussels during the Solvay Conference in October, 1930. (Courtesy of AIP Emilio Segré Visual Archives, Ehrenfest Collection.)*

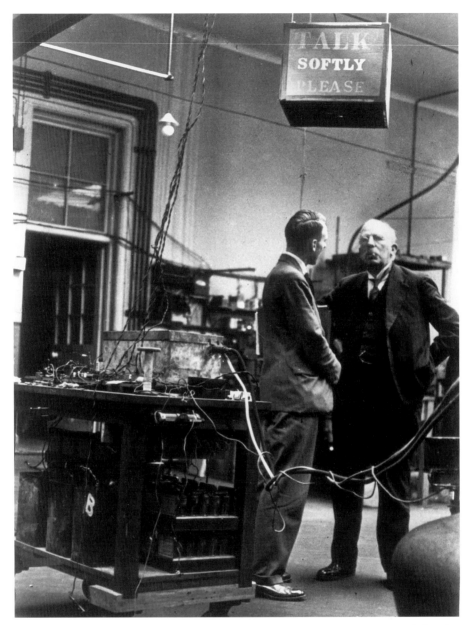

FIGURE 6   *John Ashworth Ratcliffe and Ernest Rutherford (1908) in the Cavendish Laboratory, Cambridge University on January 30, 1935. (Photograph by C.E. Wynn-Williams, courtesy of AIP Emilio Segré Visual Archives.)*

FIGURE 7   *A photograph taken by Mark Oliphant outside of the Cavendish Laboratory showing Ernest Rutherford (1908) and Niels Henrick David Bohr (1922). From the left: Mrs. Rutherford, Mrs. Oliphant, and Mrs. Bohr. (Courtesy of AIP Emilio Segré Visual Archives, Margrethe Bohr Collection.)*

FIGURE 8 *Arnold Johannes, Wilhelm Sommerfeld, and Niels Bohr (1922) in Lund, Sweden in 1919. (Courtesy of AIP Emilio Segré Visual Archives, Margrethe Bohr Collection.)*

FIGURE 9 *Niels Bohr (1922) and Max Paul Ernst Ludwig Planck (1918). (Courtesy of AIP Emilio Segré Visual Archives, Margrethe Bohr Collection.)*

FIGURE 10 *From the left: George Uhlenbeck, Henrik Kramers, and Samuel Goudsmit. (Courtesy of AIP Emilio Segré Visual Archives, Goudsmit Collection.)*

The Development of Quantum Mechanics

Werner von Heisenberg (see Figure 11) interacted with both Rutherford and Bohr during his visit to Cambridge, and out of these interactions came his formulation, in 1925, of matrix-based quantum mechanics. Schrödinger, also stimulated by the Bohr atom but working quite independently from Heisenberg, in 1926 developed the Schrödinger equation and the concept of a probability-based wave function (see Figure 11).

Paul A.M. Dirac not only showed that the Heisenberg and Schrödinger formulations were precisely equivalent but also, in his development of a relativistic equation for the electron, first postulated the existence of the positron and antimatter in general—although he was far from certain that antimatter was other than a convenient theoretical idea. This was soon to change. (See Figure 12.)

1932—Annus Mirabilis

The year 1932 was truly an *annus mirabilis* in physics, as is demonstrated in Figure 13, a composite of the first pages of now-classic papers published that year. Dirac's positron was discovered in cosmic rays by Carl Anderson; the neutron was discovered by James Chadwick; the deuteron by Harold Urey. Measurements by Kennedy and Thorndike provided experimental confirmation of the relativity of time. John Cockcroft and E.T.S. Walton produced the eponymous electrostatic accelerator and, with it, first demonstrated accelerator-induced disintegrations of atomic nuclei. Robert J. Van de Graaff developed the

FIGURE 11   *A photograph taken by Emilio Segre in 1937. From the left: an unknown individual, Victor Weisskopf, Elizabeth von Heisenberg, Niels Bohr (1922), and Werner von Heisenberg (1932). (Photograph by Emilio Segré, courtesy of AIP Emilio Segré Visual Archives.)*

first in a long series of accelerators, and E.O. Lawrence built the first cyclotron with his associate, M.S. Livingston. Figure 14 shows Carl Anderson together with Enrico Fermi, Edwin McMillen, and Hideki Yukawa at the 1952 Rochester Conference (where the author, as a nuclear physics graduate student, had the heavy responsibility of serving as chief bartender).

Figure 15 shows Petr Kapitsa and James Chadwick outside the Cavendish laboratory, obviously prepared to discover something more than neutrons! Figure 16 shows the first Cockcroft–Walton accelerator—a voltage doubler—with Cockcroft in the background and Walton allegedly in the simple cubicle at the lower right (a cubicle covered with black cloth to permit the observation of particle-induced scintillations, the detection method of choice at the time). Figure 17 shows Robert J. Van de Graaff demonstrating his first Van de Graaff and its charging belt to Karl T. Compton at Princeton. This Van de Graaff achieved something like 1.5 million volts on its spherical terminals. Figure 18 was taken at the dedication of the first of the so-called "Emperor" Van de Graaff accelerators at Yale University in 1966 and shows the speakers at the dedication ceremony. When originally installed, this Emperor accelerator could reach ten million volts on its terminal (enclosed in the pressure tank in the background of Figure 18). In the early 1980s, this tank was replaced by a much larger one, and the first extended-stretched-transuranium (ESTU) accelerator was constructed at Yale. Figure 19 is a photograph of the author inside the ESTU under the terminal, which in this accelerator can reach voltages in excess of twenty million

FIGURE 12 *Paul Adrian Maurice Dirac (1933), Kanematsu Sugiura, and J. Robert Oppenheimer in Göttingen, 1928. (Courtesy of AIP Emilio Segré Visual Archives, Uhlenbeck Collection.)*

FIGURE 13   *A collage of the first pages of a series of classic papers published in 1932.*

FIGURE 14   *The Third Annual Rochester Conference, December 19, 1952. From the left: Hideki Yukawa (1949), Edwin McMillan (1951), Carl Anderson (1936), and Enrico Fermi (1938). (Courtesy of AIP Emilio Segré Visual Archives, Marshak Collection.)*

FIGURE 15 *Petr Leonidovich Kapitsa (1978) and James Chadwick (1935) on the grounds of the Cavendish Laboratory at Cambridge University during the 1920s. (Courtesy of AIP Emilio Segré Visual Archives, Margrethe Bohr Collection.)*

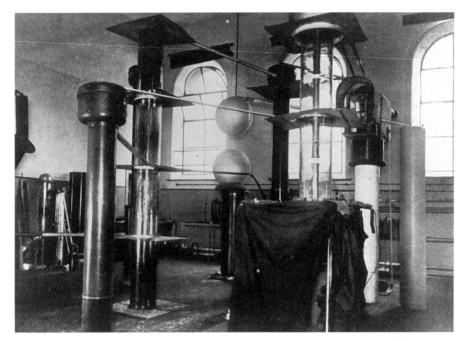

FIGURE 16   *The original Cockcroft–Walton voltage doubler in Cavendish Laboratory. The large glass tube at right with the internal segmented tubular electrodes is the acceleration tube that directed the approximately on MeV proton beam into the shielded cage shown at the lower right of the photograph. It was covered with black cloth as a light shield to permit observation of the contributions that resulted from the impact of this beam on the selected targets. The rectifiers were in the taller tube to the left, and the large spherical spark gap was included to prevent overvoltage damage to the accelerator tube. (Courtesy of Denys Wilkinson, Cambridge University.)*

volts. The terminal is surrounded by petals of the so-called "portico," which stabilizes the electrostatic gradients in the insulating gas.

Figure 20 shows the first Berkeley cyclotron, with Luis Alvarez on the extreme left and Ernest Lawrence on the extreme right. Figure 21 is a photograph of the Columbia University cyclotron, which is interesting for two reasons: first, the hand-lettered sign noting "Faculty please note, this is a CYCLOTRON"; second, on the right is a very young William Havens, who was destined to play a crucial role in the development of the American Physical Society, serving as its Deputy Executive Secretary, and then Executive Secretary, from 1956 to 1982. Figure 22 again shows the Columbia cyclotron, and the sight of I.I. Rabi cooking hot dogs on the coil head suggests that the cooling system for this cyclotron left much to be desired.

I have the impression that this sort of light-hearted attitude was much more frequent in the old days than it is now. As a further illustration of a rather less

FIGURE 17   *Professor Karl T. Compton's office at Princeton, where Robert J. Van de Graaff (left) had set up his Van de Graaff voltage generator for demonstration purposes. The rubber belt on the right column carried charge to the terminal, and Van de Graaff was judiciously using the shorting apparatus to prevent a discharge to Compton's head. (Courtesy Robert J. Van de Graaff.)*

FIGURE 18   *The Arthur Williams Wright Laboratory at Yale University at the dedication of the first Emperor, 10 MeV Van de Graaff accelerator on October 6, 1966. From the left: D. Allan Bromley, Glenn Seaborg (1951), Denys Wilkinson, Kingman Brewster, Victor Weisskopf, and Robert Van de Graaff. Unfortunately Robert Van de Graaff died shortly after this photograph was taken; this was his last public appearance.*

FIGURE 19   *The author inside the ESTU tandem Van de Graaff accelerator in the Arthur Williams Wright Laboratory at Yale University. The terminal in this accelerator can reach substantially more than 20 million volts, and the petal-like structure is comprised of a series of electrodes that stabilize the electrostatic gradient in the sulfur hexafluoride insulating gas in the pressure tank that surrounds the terminal. The targets on the tank floor are used to identify the location of sparks from the electrostatic structure to the tank wall.*

FIGURE 20   *The cyclotron in the Lawrence Berkeley Radiation Laboratory in 1944, showing from the left: Luis Walter Alvarez (1968), William Coolidge, William Brobeck, Donald Cookesy, Edwin McMillan (1951), and Ernest O. Lawrence (1939). (Lawrence Radiation Laboratory, courtesy of AIP Emilio Segré Visual Archives.)*

FIGURE 21   *The Columbia University cyclotron in the early 1940s. From the left: W. Booth, George Pegram, and William Havens. Note the sign on the upper left. Havens, a professor in the Columbia Physics Department, was to serve as Deputy and then Executive Secretary of the American Physical Society from 1956 to 1982. (William Havens, courtesy of AIP Emilio Segré Visual Archives.)*

FIGURE 22   *Again, the Columbia Cyclotron, but here with Professor I.I. Rabi (1944) cooking hot dogs on the lower coil head—indicating a lack of adequate cooling! (Courtesy of AIP Emilio Segré Archives.)*

serious approach to physics, Figure 23 shows Luis Alvarez surrounded by equipment, and although it was certainly before its time, what looks like a fiber-optic coil around his neck. Alvarez once told me that "I only did experiments that I thought would be fun," and I consider him one of the greatest experimenters of the twentieth century.

The Discovery of Nuclear Fission

Toward the close of the 1930s, Hahn and Meitner (see Figure 24) reported convincing evidence for nuclear fission. The equipment used by Hahn in this work was remarkably simple and truly "table-top" as shown in Figure 25.

The Manhattan Project

Recognition of the potential military consequences of fission came rapidly, and in the United States the Manhattan Project was established with the blessing of President Roosevelt to work toward realization of this potential. Figure 26 shows the so-called S-1 committee at its September, 1942 meeting at the Bohemian Grove in California. This committee played a major leadership role in the entire Manhattan Project.

It is not generally known that Gregory Breit, one of the giants of theoretical physics, was the first Director of the Manhattan Project. His overemphasis on secrecy—to the degree that it was seriously impacting the work of the project—was what led to his replacement by Robert Oppenheimer (see Figure 26a).

FIGURE 23   *A light moment with Luis Alvarez (1968) in the Berkeley Laboratory in the 1940s. (Courtesy of Lawrence Berkeley National Laboratory.)*

FIGURE 24   *Otto Hahn (1944) and Lise Meitner in the Hahn–Meitner Institute in Berlin in the 1950s. (Courtesy of AIP Emilio Segré Visual Archives, Physics Today Collection.)*

Figure 27 shows Vannevar Bush, who—in fact, if not in name—was the first presidential Science Advisor and, officially, the Director of the Office of Scientific Research and Development (OSRD). It would be difficult to overestimate the importance of the role Vannevar Bush played, not only during the war years—during which he provided a much needed communication bridge

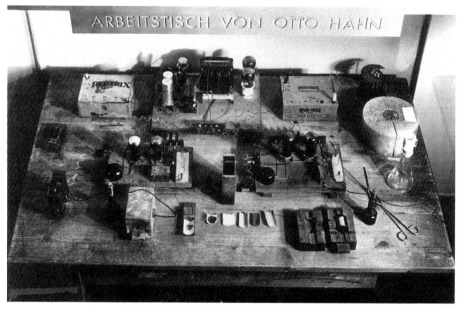

FIGURE 25 *The complete equipment used by Otto Hahn (1944) to discover nuclear fission. This was a classic table-top experiment. (Courtesy of the AIP Emilio Segre Visual Archives.)*

FIGURE 26 *A photograph taken at the Bohemian Grove in California on September 13, 1942 showing members of the S-1 Committee, which provided senior leadership for the Manhattan Project. From the left: Harold Urey (1934), Ernest Lawrence (1939), James Conant, Lyman Briggs, Edger Murray, and Arthur Compton (1927). (Ernest Orlando Lawrence Berkeley National Laboratory, courtesy of AIP Emilio Segré Visual Archives.)*

FIGURE 26a   *Robert Oppenheimer and Gregory Breit discuss Manhattan Project matters at an early Project meeting while Breit was still the Project Director. (Courtesy of Gregory Breit.)*

between the senior U.S. politicians and the scientific community—but also in the aftermath of the war, when he unofficially arranged for President Roosevelt to request a report from him making recommendations for the postwar period in science. In response to this request, Bush, largely on his own, produced the brief but remarkable document *Science, the Endless Frontier* which served as the effective blueprint for US science policy for the last half of the twentieth century. In retrospect, critics have noted that Bush used a linear model in which fundamental science led inexorably to applied science and then to technology. Critics have also noted that little mention was made in the report of any role for industry, the assumption being that, if the US public was prepared to support the research enterprise in peacetime as it had during war, then the benefits to society would be beyond imagining. The report also assumed that if federal support were channeled to the universities, then new knowledge and young minds trained to use that new knowledge in creative and innovative ways would find their way into the industrial world without any need for action on the part of government. Despite these admitted failings in the 1945 document, it has served us well. The promise has been kept. In 1998, a follow-up report appeared with the title, *Unloading our Future.* Commissioned by the Science Committee of the House of Representatives, its goal is to provide a comparable vision for U.S. science policy during the first several decade of the twenty-first century.

I have my own one-sentence definition of what our science policy should be. It is the following: *In those areas of science where our American activities do not define the frontiers, we must be working close enough to these frontiers so that we can exploit, without delay, discoveries and breakthroughs wherever and whenever made.*

The Manhattan Project, like the Apollo project, were fundamentally engineering activities, but physicists played a major role in both (particularly the Manhattan Project).

Figure 28 shows Lawrence, Fermi, and Rabi discussing the Manhattan Project during its early days at Los Alamos, and Figure 29 shows Niels Bohr—or "Mr. Baker," as he was called in Los Alamos—with the Los Alamos Director, J. Robert Oppenheimer. This figure illustrates the close cooperation that existed between the remaining free scientific community in Europe and the Americans, but it also gives some indication of the heavy responsibilities that rested on Oppenheimer's shoulders (compare this photograph with Figure 12, showing Oppenheimer in happier days).

Figure 30 shows Enrico Fermi receiving a richly deserved Medal of Merit from General Groves, the military commander of the Manhattan Project.

FIGURE 27   *Vannevar Bush in his laboratory. (Copyright NBS Archives.)*

FIGURE 28　*Ernest Lawrence (1939), Enrico Fermi (1939), and I.I. Rabi (1944) in the early days at Los Alamos. (Ernest Orlando Lawrence Berkeley National Laboratory, courtesy of AIP Emilio Segré Visual Archives.)*

Figure 31 is a rather rare photograph which shows Edward Teller congratulating Robert Oppenheimer after he received the Fermi Award from President Johnson. This symbolic reconciliation in many ways marked the end of a very unhappy period during which Teller and Oppenheimer disagreed strongly about the wisdom of developing the hydrogen bomb, during which Teller negotiated the creation of his own weapons laboratory at Livermore, and during which the Oppenheimer Hearings recommended the removal of Oppenheimer's security clearances. The years of stress had clearly left their mark by the time this photograph was taken.

Figure 32 shows one of the mushroom clouds that have become emblematic of nuclear science. This one is the result of the most recent 1992 series of tests by the French government in the South Pacific. The decision to conduct

FIGURE 29   *Niels Bohr (1922) and J. Robert Oppenheimer in discussion at Los Alamos in the early 1940s. (Niels Bohr Archive, courtesy of AIP Emilio Segré Visual Archives.)*

FIGURE 30   *General Groves pins the Medal of Merit on Enrico Fermi (1938). The other scientists in attendance at this ceremony had previously received this same Medal. They are, from the left, Harold Urey (1934), Samuel Allison, Cyril Smith, and Robert S. Stone. (Courtesy of Laura Fermi.)*

FIGURE 31   *On December 2, 1963, exactly ten years after he was charged with being a security risk, J. Robert Oppenheimer received the Fermi Award from President Lyndon Johnson. After the White House ceremony, Edward Teller, who had won the same award in 1962, warmly congratulated his long-time colleague. Teller's role in the Oppenheimer hearings and the differences between them, particularly with reference to the development of the H bomb, had much to do with the establishment of the Lawrence Livermore National Laboratory. This is a rather rare photograph showing how Oppenheimer had aged over the preceding decade. (Photo by Ralph Morse, courtesy of TimePix.)*

FIGURE 32 *One of the French tests of an atomic weapon in the South Pacific. This test was conducted in 1995 on the island of Mururoa. (Courtesy of Commissariat à l'Énergie Atomique, France.)*

these tests was unfortunate for several reasons: international politics, radioactive fall-out, and, perhaps most important of all, the reinforcement of a public aversion to anything associated with nuclear science.

The MIT Radiation Laboratory

The other major wartime laboratory, the Radiation Laboratory at MIT, was devoted to the development of radar. Figure 33 shows a typical laboratory during the early days of this project.

FIGURE 33   *The MIT Radiation Laboratory in the early 1940s. (Courtesy of Massachusetts Institute of Technology.)*

The Merger of Natural Philosophy and Invention

In both the Manhattan Project and the Radiation Laboratory activity (as well as in what is often forgotten—namely, the major work being carried on in the nation's medical schools and hospitals toward providing adequate medical care on the battlefield) each group of scientists had very clear objectives and no one much cared what specific scientific or technical background any individual brought to their project. What counted was whether the participants were intelligent and whether they were prepared to devote their full effort toward the goals at hand. These activities ushered in a seachange in the scientific and technical communities. Prior to the war basic research was, to a large extent, directed toward the *understanding* of nature, while invention and technology were directed toward the *mastery* of nature. Research thus took a parallel and insular course. What the wartime projects demonstrated was that basic understanding could greatly facilitate the development of technology and basic technology could facilitate whole new areas of basic research. The prewar activities that had frequently been called "natural philosophy" and "invention," respectively, were irretrievably joined, and nowhere more so than in physics. Physics departments in universities across the nation received freight car loads of war surplus equipment and literally thousands of highly trained scientists and engineers, many of them as a consequence of the GI Bill—one of the best pieces of legislation ever passed by a U.S. Congress.

Physics in the 1930s

Let us now shift our attention back to physics and the natural world and away from physics and science policy. Two of the most important developments of the century, quantum mechanics and relativity theory, had already been established by mid-century, and members of the scientific community had been waiting throughout the war years for a chance to apply them in peacetime research.

In the 1930s, Felix Bloch recognized that as atoms were brought close together (as in solids), the discrete energy levels that characterized them when they were well-isolated in a gaseous phase spread, resulting in electron bands and, even more important, forbidden bands. This discovery immediately made it possible to understand, at least in broad outline, the difference between monovalent metals, divalent metals, semiconductors, and insulators—all as a function of the size of the energy gap that separated filled valence bands from the open conduction bands. These differences are shown in very schematic fashion in Figure 34 and in somewhat more detail in Figure 35. In 1940, Frederick Seitz pulled together what was known about solids in a classic textbook entitled *The Modern Theory of Solids,* which became required reading for generations of students as well as more senior graduates of the wartime projects. Figure 36 shows Seitz in 1937. In addition to his leadership in solid-state physics (or, condensed matter physics, as it is now called), Fred Seitz went on to serve as President of the U.S. National Academy of Sciences and, subsequently, President of the Rockefeller University.

The Immediate Postwar Period

The immediate postwar period was one of major excitement as graduates of the wartime projects began to focus on where their subsequent careers would lead. A catalyst in this process was a series of symposia and conferences that brought physicists together to exchange ideas. The Shelter Island Conference of 1947 was typical of these meetings (see Figure 37). So also were the series of meetings held at the Carnegie Institution in Washington (see Figure 37a). Most of the leading figures in American physics could still be photographed standing together in an entryway!

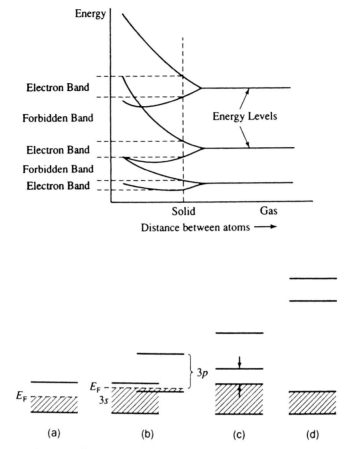

FIGURE 34   *A schematic illustration of electron energy levels in a gas and in various solids. The lower panel shows the situation in a monovalent metal (a); in a divalent metal (b); in a semiconductor (c); and in an insulator (d). E_F represents the Fermi level.*

PART II
The Explosive Growth of Postwar Physics, 1950–1999

Introduction

The number of such meetings increased rapidly as physicists from the Manhattan Project, the Radiation Laboratory and other major wartime facilities, who had dispersed to the universities and to a few industrial organizations, came together to try to obtain an overview that would permit them to select a particular area of research that they felt was most promising and best suited to their talents and equipment.

Train loads of equipment declared surplus in the governmental laboratories were arriving at many of the nation's universities, where an explosion of new experimentation became possible as a result of technological developments during the war years.

In relatively short order, individual university groups selected their areas of concentration, attracted graduate students in large numbers from the cadre of individuals who had interrupted their studies for military service, and began their research programs.

The availability of federal funding orders of magnitude greater than any available in the prewar years was also a major factor. Funds were channeled through agencies such as the Office of Naval Research, the Atomic Energy Commission, and the National Science Foudation—the latter two established in 1946 and 1980, respectively.

In parallel, the GI Bill of 1944—one of the most important single legislative actions of the century—opened the doors of universities and colleges, for the first time, to many for whom higher education had previously seemed economically impossible.

The availability of equipment, experienced faculty members, and highly motivated students fueled the exponential growth of American physics research—both experimental and theoretical. In a relatively short period, the generous investments made by Congress, and ultimately by the U.S. taxpayers, led to a scientific and technological renaissance unlike anything the world had seen before.

I now turn to some of the specific areas of growth in postwar physics.

Materials Science

The behavior of solids and of fluids began to occupy what is now, by a substantial margin, the largest subdivision of physics. This reflected (in part, at least) the enormous importance of the applications made possible by new understanding in material science and in condensed matter generally. In Figure 38, I list seven subareas of material science, ranging from superpure materials to superconductors, where major progress has been made.

Today it is possible to obtain pure materials at the level of one part in 10^{12}, and (for extra cost) one part in 10^{14}. It was the availability of increasingly pure silicon since 1960, for example, that improved the transparency of optical fibers by something like a hundred orders of magnitude. We now have materials—often organic polymers—with a strength to weight ratio at least 50 times better than was available at the end of World War II. In the realm of electronic materials, the basic research has thus far been successful in keeping ahead of Moore's Law—the statement that the number of active elements on an integrated circuit chip doubles every 18 months and has done so for more than the

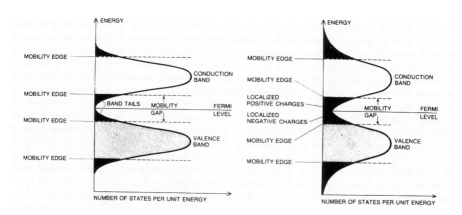

FIGURE 35   *Amorphous semiconductors are not strongly disordered (left), and have valence and conduction bands similar to those in the corresponding crystalline semiconductor. The distinguishing feature of the bands in amorphous solids is the replacement of the sharp band edges present in crystals by what are called "band tails" or localized states that extend into the energy gap. The localized states are separated from the extended states in the main part of the bands by mobility edges. The region that lies between the mobility edges of the valence and conduction bands is the so-called mobility gap. It plays the same role in amorphous semiconductors that the energy gap plays in crystalline semiconductors. Chemical impurities are defects in the configuration of local bonds and can lead to sharp structural changes (not shown) in the "mobility gap." If the disorder is large, as is expected in multicomponent glasses (right), the band tails of the valence and conduction bands can overlap in the mobility gap. This leads to a redistribution of electric charge as electrons move from one localized state to another in order to lower their energy. The result is a high density of positively and negatively charged traps which decrease the mobility of the carriers and make the material less sensitive to efforts to modulate its conductivity by chemical means—that is, by doping. (Courtesy of David Adler, MIT.)*

FIGURE 36   *Frederick Seitz in 1937. (Courtesy of AIP Emilio Segrè Visual Archives, Physics Today Collection.)*

FIGURE 37 *At the Shelter Island Conference in 1947. From the left: Willis Lamb (1955), Abraham Pais, John Wheeler, Richard Feynman (1965), Herman Feshbach, and Julian Schwinger (1965). (Courtesy of the AIP Emilio Segre Visual Archives.)*

past two decades. Recent developments have made it clear that we can expect Moore's Law to be valid for at least the next five to seven years.

Our ability to understand, to probe, and to structure surfaces has opened up entirely new areas of study in catalysis and corrosion resistance, and an entirely new understanding of phenomena such as friction and adhesion. The development of nanotechnology—which makes it possible to create devices from silicon crystal substrates, for example, that contain not a single dislocation—has led to improvement in the sensitivity of accelerometers and sensors generally by many orders of magnitude. Entire optical benches and chemical laboratories are now being fabricated on single chips with nanoscale rotary and linear motors powering the necessary motions. The ability to craft sensors, together with their associated arithmatic units and transmitters to link them to external central processors, opens up entire new fields of activity in mechanics and medicine, in biology, and in fields we have not yet even imagined. The development of new materials has also had a major impact on our ability to develop human prosthetic devices that will replace both bones and soft tissue or serve as matrices for the regrowth of either. We have learned how to reduce corrosion on prosthetic device surfaces by factors of many hundreds and, in consequence, have extended their lifetimes after implantation in the human body to more than an average human lifetime. The day of the bionic man or woman is closer than we may think.

I have included Figure 39, which shows lead–tin telluride crystals growing on a substrate, because I find it intrinsically beautiful and because it demonstrates in striking fashion the diversity that exists in such a simple and commonplace process.

Superconductivity

And, finally, there is the phenomenon of superconductivity. After almost a century of slow and frustratingly small increases in the critical temperature at which materials become superconducting, we have made a major increase in that critical temperature through the use of ceramic superconductors—although we have still not reached the "Holy Grail" in this field, namely, a material that is

FIGURE 37a *Participants in a symposium on physics held at the Carnegie Institution in Washington circa 1949. In the front row, from the left, are Fleming, Bartlett, Thomas, Bethe (1967), Bohr (1922), and Bloch (1952); in the second row, from the left, are Wigner (1963), Plesset, Kalckar, Wheeler, and Gamow; in the third row, from the left, are Tuve, Heydenburg, Meyer, Franck (1922), Roberts, Furry, Hafstad, Teller, Crane, Seeger, Breit, and Critchfield. (Niels Bohr Archive, courtesy of AIP Emilio Segré Visual Archives.)*

superconducting at room temperature. Because of improvements in manufacture and fabrication of these materials, superconductivity is just beginning to move from the domain of the research magnet into motors, generators, and other components in use throughout society, where even small increases in efficiency can add up to important energy savings. Superconductivity is also finding application in high-capacity transmission lines and transformers, where there are major qualitative as well as quantitative advantages, and it is finding other novel uses such as the suspension of magnetically levitated trains. Superconductivity has long been a technology waiting for application, and the indications are that we are on the threshold of major change in this area.

Superconductivity was originally discovered in 1911 in mercury at about 4 K by Kammerlingh Onnes in Leiden (he is shown with his original equipment in Figure 40). Figure 41 plots the transition, or critical temperature, as a function of time beginning in 1911 and shows the sharp break that occurred in 1986 with the introduction of a entirely new class of ceramic materials. The mechanism whereby materials lost all electrical resistance had been a puzzle for decades until John Bardeen, Leon Cooper, and Robert Schrieffer succeeded in developing a theory that explained superconductivity in metallic samples. Central to their theory was the role of Cooper pairs—pairs of electrons coupled with opposing spins and angular momentum, which in consequence were able to move freely through the metallic structure. Figure 42 shows the trio at their Nobel Prize ceremony in 1972.

Figure 43 shows Muller and Bednorz in their IBM Zurich laboratory following the 1987 announcement that they were to be awarded the Nobel Prize for their development of ceramic superconductors. Figure 44 is an artist's conception of the

MATERIALS & SCIENCE

1. **Superpure Materials**
   Input to fiber optic cable 99.9999% purity
   Transparency increase by 100 orders of magnitude since 1960

2. **Superstrong Materials**
   Strength increases by factor of 50 since 1900
   In 1900 to support 25 tons required a cast iron bar 1 inch square weighing 4 lbs/foot; today a polymer fiber 1/3 of an inch square weighing one ounce/foot

3. **Electronic Materials**
   Replacement of silicon dioxide with silicon nitride as the insulator in integrated circuits will extend Moore's Law by at least 5 years

4. **Surfaces**
   Visualization and control of surfaces at the atomic and molecular level allow major progress in understanding and improving corrosion, wear, friction, color, catalytic behavior, and the like

5. **Sensors**
   Use of MEMS technology allows fabrication of entire sensor and associated chemical and electronic components on a single chip with vastly improved lifetimes, reliability, and calibration stability

6. **Biomaterials**
   We are on the threshold of being able to implant at least crude artificial retinas, and artificial semi-cochlear canals and connect them to the brain to provide some level of sensory perception to the blind and to the deaf. New biomaterials can serve as templates for regrowth of both hard and soft human tissue

7. **Superconductors**
   Again we are on the threshold of widespread application of superconductors in motors, generators, transmission lines, maglev transport, and the like. The Holy Grail here is the room-temperature superconductor, which still eludes us.

FIGURE 38

FIGURE 39   *Lead-tin telluride crystals growing on a substrate. Magnification here is times 60. (Courtesy Plessey, Ltd.)*

planar structure of a typical ceramic superconductor. Unfortunately, there is not, as yet, any generally accepted theory of the superconductivity in these structures equivalent to that developed by Bardeen, Cooper, and Schrieffer for metallic samples. As is evident in Figure 41, minor adjustments in the composition and processing of these ceramics has increased the critical temperature to 160 K. But progress has slowed dramatically, and it is not yet clear whether it is possible, even in theory, to reach room-temperature (300 K) superconductivity.

FIGURE 40   *H. Kammerlingh Onnes (1913) in his laboratory at the University of Leiden in 1911. (Courtesy of AIP Emilio Segré Visual Archives, W.F. Meggers Collection.)*

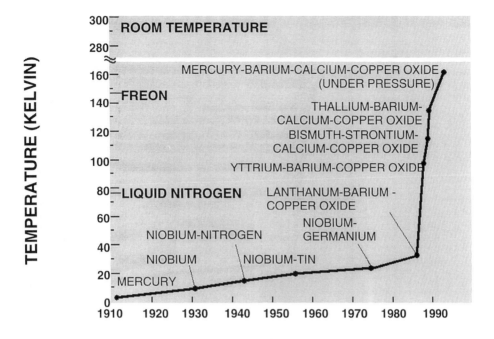

FIGURE 41 *Critical temperatures for various superconducting materials plotted as a function of time. The break in the ordinate indicates that we have a long way to go before reaching the "Holy Grail" of a room temperature superconductor.*

FIGURE 42 *From the left: John Bardeen (1956, 1972), Leon N. Cooper (1972), and John Robert Schrieffer (1972). (Courtesy of AIP Emilio Segré Visual Archives.)*

FIGURE 43   *Bednorz (right) and Müller after winning the Nobel Prize for Physics in 1987 for their discovery of ceramic superconductors. (Courtesy of IBM Zurich Research Laboratories.)*

Buckyballs and Nanotubes

Happily, surprises continue to occur. In 1996, Richard Smalley of Rice University received the Nobel Prize for his discovery that in electrical arcs between carbon electrodes entirely new all-carbon molecular complexes were being formed. Among them was a 60 atom molecule, which took a form analogous to a giant soccer ball or a dome constructed by Buckminster Fuller hence, the generic name of "fullerene". Subsequently, it has been shown that there are a number of structural siblings, the smallest of which has 36 carbon atoms in its structure. These molecules can now be made in quantity and can be used to carry other materials or molecules enclosed within them for drug delivery or other specialized use. There is also evidence that layers of the fullerenes can be superconducting. More recently, Smalley and his collaborators demonstrated that not only it is possible to produced fullerenes, but also cylindrical tubes of nanometer dimension and arbitrary length—again, in quantity. These have very interesting mechanical and electronic properties and also have the possibility of being superconducting, and they are enormously strong in the mechanical sense. Figure 45 shows Smalley in his laboratory, together with schematics of both the fullerenes and the nanotubes. The different structures have quite different color characteristics when in solution.

Surface Science

One of the most important developments in surface science during the past century was the invention, in 1981, of the scanning tunneling microscope (STM). Figure 46 shows Heinrich Rohrer and Gerd Binnig of the IBM Zurich Research Laboratories working with one of the prototype devices. The STM, for which they received the 1986 Nobel Prize in Physics with Ernst Ruska, functions by scanning an atomic scale tip over a conducting surface and measuring the tunneling current to the tip to provide a profile of the surface at the atomic level. This invention was followed in 1985 by the advent of the atomic force microscope (AFM) by Binnig together with Christof Berber of IBM Zurich and Calvin Quate of Stanford University. In contrast with the STM, the AFM images nonconducting surfaces, including living cells at the atomic level. The AFMs and STMs have the additional advantage of being able to pick up atoms and move them relative to the surface being studied. As Richard Feynman once put it, "the physicists can now put atoms where the chemists want them." Figure 47 is a striking example of what can be done with this technology. This shows what its authors have dubbed a "quantum corral," consisting of 48 iron atoms absorbed onto a copper surface so that it encloses ripples in the local density of surface electron states.

FIGURE 44   *An artistic representation of a ceramic superconductor composed of mercury, barium, calcium, copper, and oxygen that becomes superconducting at 133K. (Courtesy of Judy Franz, American Physical Society.)*

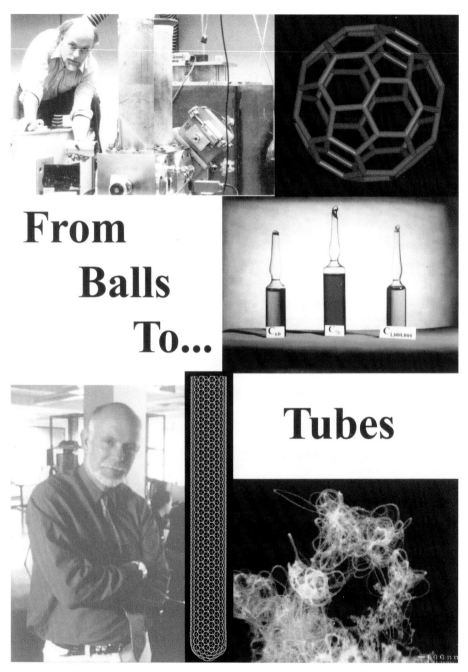

From Balls To...

Tubes

FIGURE 45   *Clockwise from the upper left: Richard Smalley (1996) working on the equipment in 1985 on which buckyballs were discovered, a computer graphic of a C_{60} buckyball, three vials of different fullerenes in solution (the first two are C_{60} and C_{70} in toluene and the last is single-wall nanotubes labelled here for effect $C_{one\ million}$ in acetone), a scanning electron micrograph of a raw single-wall nanotube material made by the Rice Laser Vaporization method, a computer graphic of a single-wall nanotube (here the so-called 10/10 tube which is to nanotubes what C_{60} is to fullerenes) and, finally, a recent photo of Richard Smalley in the Nanoinstrumentation Facility of the Rice University Center for Nanoscale Science and Technology. (Courtesy of Richard Smalley, Rice University.)*

Fluid Physics

As the name implies, condensed matter includes fluids as well as solids; here, again, substantial progress has been made in understanding not only the structure of liquids but also the physics involved in the various flow patterns, ranging from laminar to turbulent, that appear in just about every physical situation involving fluids. Figure 46 illustrates the transition from laminar flow on the left to turbulent flow on the right, while the center panel shows intricate but highly ordered flow that can occur between those two limiting cases, illustrating the apparently contradictory but still true fact that in chaos there is order.

During the 1950s, Edward Lorentz at MIT was continuing in his attempt to understand weather prediction using the rather rudimentary computers available to him. Prior to about 1960, it had been an article of faith among scientists in many fields that if one knew the governing equations for the physical system that one wished to study, and used a particular set of measurements of the pertinent parameters as input, then a phenomenon known roughly as *convergence* was understood to hold—i.e., if the input parameters varied slightly, the predictions would be stable and small input perturbations would not affect at least the gross features of the predictions.

In 1961, however, Lorentz discovered in his meteorological studies—more

FIGURE 46   *The scanning tunneling microscope's inventors, Gerd Binnig (right) and Heinrich Rohrer of the IBM Zurich Research Laboratory, with one of their early devices. Their design of the STM earned them the 1986 Nobel Prize in Physics. (Courtesy of IBM Zurich Research Laboratories.)*

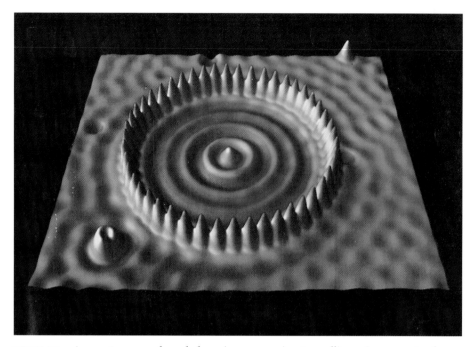

FIGURE 47   *A quantum corral made by using a scanning tunnelling microscope to place 48 iron atoms in the 143 angstrom diameter circle on a {111}Cu surface. The same STM was then used to image the ripples in the surface electron local density states; these ripples closely match what would be expected for a particle confined in a circular box. Some of the electrons end up in a two-dimensional electron gas bounded by the work function on the vacuum side and by a band gap in the energy spectrum of the bulk electrons on the copper side. (Courtesy of IBM Almaden Research Center.)*

or less by accident—that small changes in the input parameters, as long as the governing equations had nonlinear terms, could result in large and dramatic changes in the predictions of his model. He coined the name "butterfly effect" to describe this, recognizing that the fluttering wings of a Brazilian butterfly could, in principle, set off striking changes in weather thousands of miles away.

Attempting to simplify the physical system under study, Lorentz turned to thermal convection in a fluid. He stripped down the Navier–Stokes equations, (which traditionally have been used to describe fluid motion), leaving only a three-state model but retaining nonlinearity in this model. At any given moment, the state of such a system is described by a single point in the three-dimensional diagram of Figure 49.

What Lorentz found was that this phase point moved in a sort of two-winged loop but never actually retraced its exact path—and, indeed, never crossed or intersected that path. The phase point, however, appeared to follow a general path that became known as the Lorentz attractor. The understanding of chaotic phenomena and their dependence on both initial conditions and

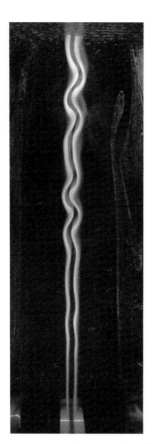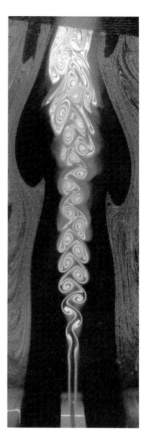

FIGURE 48 *Photographs of fluid flow pattern. On the left of the flow is laminar and on the right, turbulent. The middle panel shows some of the intricate structure in the intermediate scale. (Courtesy of Judy Franz, American Physical Society.)*

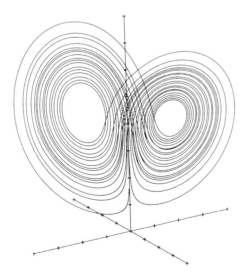

FIGURE 49 *An illustration of the Lorentz attractor for a dynamical system. The phase point over time traces out this characteristic "owl mask" but without ever closing. The Lorentz attractor has become something of an emblem for chaotic systems. (Courtesy of Paul Bourke, Swinburne University of Technology.)*

nonlinearity that evolved from Lorentz's work marks one of the major achievements in twentieth-century physics.

Self-Similarity

One other important development in the study of chaotic (or apparently chaotic) systems was initiated in the early 1960s by Benoit Mandelbrot, then at IBM. He had noted that when cotton prices were plotted as a function of time,

FIGURE 49a   *The Koch Curve looks the same at all levels of magnification, and it therefore illustrates the phenomenon of self-similarity. The concept of self-similarity, particularly as it was developed by Benoit Mandelbrot, has found application throughout the physical sciences as an effective representation of such natural formations as shorelines and mountain profiles, and (more recently) in economics, where it has been used to represent corporate stock prices over time (Courtesy Benoit Mandelbrot, Yale University and IBM.)*

the fluctuations appeared to be chaotic. But while each particular price change was random and unpredictable, the sequence of changes was independent of scale—plots of daily fluctuations matched those for monthly or annual fluctuations. Moving beyond his economic model, Mandelbrot discovered the self-similarity phenomena in a wide variety of systems. To describe what he was seeing, he introduced the concept of fractional dimensions, and in 1975 he coined the word "fractal." Mandelbrot has said that, "a fractal is a way of seeing infinity."

This idea is illustrated in Figure 49a, an image of the so-called Koch curve. It is constructed by drawing an equilateral triangle with sides of unit length. On the middle of each side of the triangle a new triangle one third the size is added and this process continues. The length of the boundary of the Koch curve $3 + 4/3 + 4/3 + 4/3 + \ldots$ infinity, but at the same time the area bounded by this infinite perimeter is less than that of the circle that circumbscribes the original triangle.

The availability of modern computers has made possible major extension of these concepts; an example is shown in Figure 49b.

FIGURE 49b   *An example of self-similarity. The large figure is a magnification of the area within the white rectangular outline in the inset figure. Remarkably beautiful figures may be produced using Mandelbrot's fractal geometry, and the concept has increasingly found its way into the realm of the visual arts. (Courtesy Benoit Mandelbrot, Yale University and IBM.)*

Development of the Transistor

In the immediate postwar period, physicists coming back to universities were bubbling with enthusiasm and full of new ideas that they had not been able to follow up during the war years; the whole field of physics seemed, in a very real sense, to be "riding off in all directions." This, however, was not universally true. In the Bell Laboratories, for example, where some of the nation's most able physicists had come together in the late 1940s, the laboratory management insisted on a more targeted pattern of activity and established a number of specific challenges toward whose solution they expected the laboratory staff to direct their attention. One of these challenges was that of finding a replacement for the notoriously unreliable and short-lived vacuum tube. The answer, of course, was the transistor, invented in 1947. Figure 50 is a photograph of the first transistor, and Figure 51 shows its inventors, John Bardeen, William Schockley, and Walter Brattain. The transistor had three obvious advantages over the vacuum tube: It could obviously be made much more compact, it used much less energy, and it had effectively an infinite lifetime. These were the essential characteristics of the computer of the future. The inventors of the device that made the digital computer possible received the 1956 Nobel Prize in Physics.

The Evolution of Computers

During the war years, the group at Los Alamos carried forward the whole field of computation and physical modeling beyond anything that had previously been imagined, but it was recognized that the limitations imposed by vacuum tubes clearly established physical ceilings on the power and scope of future computers and computation. Three of the individuals most involved in the Los Alamos computing—Stanley Ulam, Richard Feynman, and John von Neuman—are shown in Figure 52, in an old Los Alamos photograph of the trio discussing the future of computing. Figure 53 shows the ENIAC computer at the University of Pennsylvania, which in 1946 defined the state-of-the-art. The superimposed operator of the laptop is intended to demonstrate how technology had changed primarily because of the invention of integrated circuits by Robert Noyce and Jack Kirby. As an illustration of how much change has occurred, Figure 54 shows, on precisely the same scale, a recent Hewlett–Packard microprocessor containing 4.5 million transistors with their associated resistors and capacitors juxtaposed with an American nickel.

Figure 55 illustrates what has come to be known as Moore's Law, after Gordon Moore's observation that for several decades the number of active transistor elements per chip doubled every 18 months. This figure is not up-to-date,

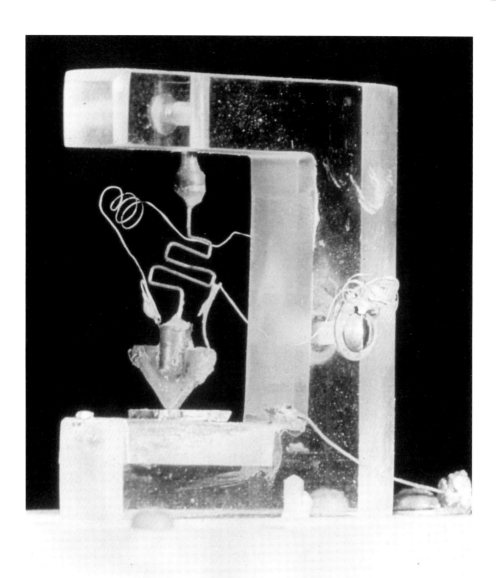

FIGURE 50 *The world's first transistor, developed in the Bell Telephone Laboratories in 1947. (Courtesy Bell Telephone Laboratories.)*

and the latest information indicates that Moore's Law can be expected to remain valid at least until between 2005 and 2007. In 1998, it appeared that the flattening off shown in the shaded area was in fact at hand as electron tunneling through insulating layers of silicon dioxide on the chips appeared to pose a real physical limit to Moore's Law's continued validity. T. P. Ma at Yale recently demonstrated that replacement of the silicon dioxide by silicon nitride

FIGURE 51   *From the left: John Bardeen (1956, 1972), William Shockley (1956), and Walter H. Brattain (1956), the inventors of the transistor. (Courtesy Bell Telephone Laboratories.)*

FIGURE 52 *From the left: Stanislaw Ulam, Richard Phillips Feynman (1965), and John von Neumann discussing the future of computing at Los Alamos in 1947. (Courtesy of AIP Emilio Segré Visual Archives, Ulam Collection.)*

FIGURE 53 *The ENIAC computer at the University of Pennsylvania in 1946. A chance meeting between John von Neumann and Konrad Zuse of Germany led to von Neumann's establishment of a group at the University of Pennsylvania charged with creating the ENIAC (Electronic Numerical Integrator and Computer). The ENIAC was designed by J. Prosper Eckert and John Mauchly. They used 18,000 vacuum tubes at a cost of $400,000; it is widely considered to be the progenitor of modern computers. The Newsweek editors superposed the figure on the right with her laptop to emphasize the enormous contrast in size and power that has taken place since 1946 in the world of computers. (Courtesy of Newsweek.)*

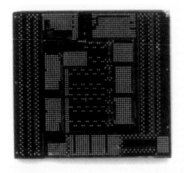

FIGURE 54 *A state-of-the-art Hewlett–Packard microprocessor chip containing 4.5 million active transistor elements with their associated resistors and capacitors compared in size to an American nickel.*

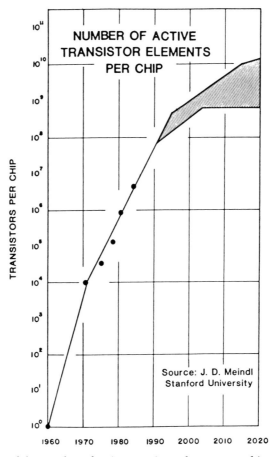

FIGURE 55 *A plot of the number of active transistor elements per chip against year—illustrating what is known as Moore's Law. (Courtesy J.D. Meindl of Stanford University.)*

with quite different dielectric characteristics promises to extend validity for at least five to seven years.

Breakthroughs in Communication

A doubling time of 18 months for the number of active elements per chip is clearly remarkable, but, as Figure 56 demonstrates, single optical fiber transmission bandwidths have been doubling every nine months and the actual in-the-field telephone company products now lag behind the frontiers of research by only four years. The resulting communication and computation explosion has truly created a global village.

Arguably, one of the most important single developments in making this additional scope and power accessible and much more user-friendly was the advent of the World Wide Web, introduced by Tim Berners-Lee at CERN (see Figure 57). Berners-Lee is an Oxford-educated physicist-turned-programmer in the CERN High Energy Physics group. Much of his communication with other high-energy physics centers was via the Internet, which had been assembled from various proprietary agency networks within the US government; these included ArpaNet, DOENet, NSFNet, and the like. CERN made its first connec-

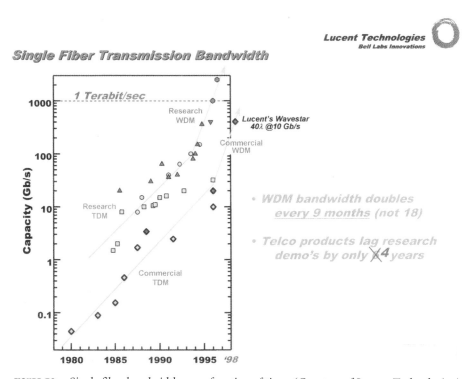

FIGURE 56   *Single fiber bandwidths as a function of time. (Courtesy of Lucent Technologies.)*

FIGURE 57   *Former physicist Tim Berners-Lee invented the World-Wide Web as an essential tool for High Energy Physics (HEP) at CERN from 1989–1994. (Courtesy of CERN.)*

tion to the Internet in 1989, and by 1990 it was the largest European Internet site and a major force in the further development of Internet capabilities both in Europe and throughout the world. As someone very much interested in CERN's future, Berners-Lee in 1989 proposed the creation of the World Wide Web and by 1991 he had developed the first effective browser.

No one did, or could have, predicted the consequences. Prior to April 1993 the Internet was restricted to US governmental or governmentally sponsored traffic, but at that time, President Clinton opened the Internet to the entire world and it has subsequently penetrated to the most remote corners of our planet. From a parochial point of view, it has also changed entirely the nature and scope of communications within the world of physics. Physicists now routinely carry out experiments remotely with both control and experimental data flowing over the Internet.

A development just over the horizon is the quantum computer. I was unable to find a figure that I felt in any way made clear just how a quantum computer would function if and when one is built; so instead, in Figure 58, I quote one of the often repeated goals that is believed to be applicable to at least the class of programs that includes the factoring of large prime numbers, the searching of very large databases, and the like. It is a startling prediction that a functioning quantum computer could, in 30 seconds, achieve results that would require 10 billion years on the best of our present supercomputers! Developing a quantum computer is surely the grandest of all grand

IN THEORY, A QUANTUM computer could solve, in

about 30 seconds, problems that would occupy a

conventional supercomputer FOR 10 BILLION YEARS.

FIGURE 58

challenges, and physicists around the world have accepted this challenge. I firmly believe that they will succeed within the decade and that when they succeed it will open up an entirely new era of physics and of the modeling of physical systems comparable to the introduction of solid-state computers in the 1960s.

Computational Chemistry

One of the areas already being transformed by the availability of high-speed, high-capacity computers is computational chemistry. Two examples that I show in Figures 59 and 61 were obtained from my colleagues, William Jorgenson and

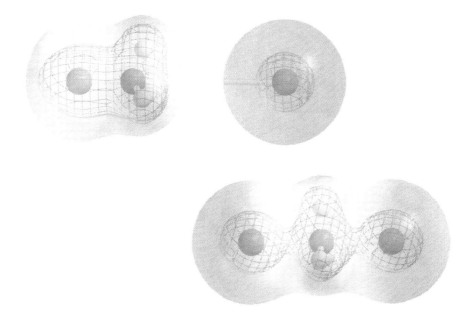

FIGURE 59 *Computer modelling of the interaction of the negative chlorine ion with a methyl chloride molecule calculated using a Gaussian 80 program. The structures and charges thus obtained were used to calculate the effective hydration by umbrella sampling and free energy perturbation methods in Monte Carlo calculations. Each structure is surrounded by a grid representing the electrostatic potential and an isoelectronic surface colored according to the electrostatic potential. (Courtesy William Jorgenson and Julian Tirado-Rives, Yale University.)*

Julian Tirado-Rives of the Yale chemistry department. Figure 59 shows the transition state structure at different chlorine to carbon distances for the reaction $Cl^- + CH_3Cl \rightarrow ClCH_3 + Cl^-$ calculated using the Gaussian 80 program. The structure and charges thus obtained were used to calculate the effect of hydration by umbrella sampling and free energy perturbation methods in a Monte Carlo calculation. Each structure is surrounded by a grid representing the electrostatic potential and an isoelectronic surface colored according to the electrostatic potential. This, in effect, is dry chemistry, where the entire chemical process can be followed in detail, allowing entirely new insight into the electronic behavior involved.

Folding of Proteins

One of the major challenges in modern biochemistry and biophysics has to do with the folding of proteins, as illustrated schematically in Figure 60. Having assembled the amino acids in the appropriate order, according to information stored in the cell DNA, the functional protein must be folded into a very specific shape and in a very specific way. How this comes about is still, to a large extent, a mystery. But progress is being made. One of the methods used in this work is illustrated in Figure 61, where the molecular dynamics of the unfolding of the protein apomyoglobin is simulated as it would occur in water over a period of 350 picoseconds at a temperature of 298 K. Because there were no experimental structures known for apoymoglobin, all of these simulations were started from the ex-

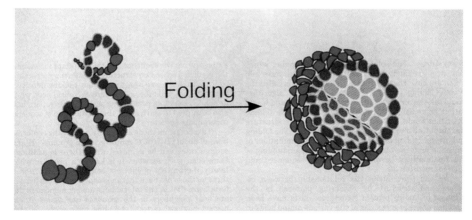

FIGURE 60 *The folding of a globular protein from its denatured state (left) to its unique compact native state (right) is one of the most important problems in modern biology and is often referred to as the "second genetic code." The folding pattern is encoded in its sequence of amino acid monomers but the folding code remains a mystery. (Courtesy Frederick Richards, Yale.)*

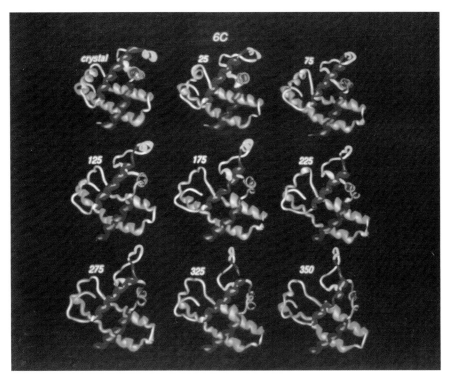

FIGURE 61 *A molecular dynamic simulation of the unfolding of apomyoglobin conducted in water for 350 picoseconds at 298K. Since there are no current experimental structure data available for apomyoglobin all simulations were started from the experimentally determined crystal structure of sperm whale myoglobin after removal of the heme submolecule. (Courtesy of Julian Tirado-Rives and William Jorgenson, Yale University.)*

perimentally determined crystal structure of sperm whale myoglobin after removal of the heme component. From simulations such as this, we have been able to assemble an increasingly detailed knowledge of the dynamics of protein folding.

Photon Probes

Much of the experimental information concerning chemical dynamics is being obtained through study of the interaction of photons of wavelengths ranging from infrared to X-rays, and particularly in very fast pulsed-photon probed experiments; the laser is one of the work horses in such studies. Figure 62, a photograph taken in 1965, shows Nikolai Basov, Charles H. Townes, and Alexander Prokhorov, the winners of the 1964 Nobel Prize for Physics for their parallel development of the laser.

In recent years, the development of ever-brighter X-ray sources has made possible the use of Laue diffraction to unravel the structure of even very complex biological molecules. Figure 63 shows the average spectral brightness as a

FIGURE 62   *The winners of the Nobel Prize for Physics in 1964. This photograph taken in 1965 shows, from the left, Nikolai Gennadievich Basov, Charles Hard Townes, and Alexander Mikhailovich Prokhorov. (Sources for the History of Lasers, courtesy of AIP Emilio Segré Visual Archives.)*

function of time during this past century—from a variety of sources ranging from candles to third generation undulators in electron synchrotrons. The increase from the early 1960s to the present has been obviously dramatic. Figure 64 shows some of the beam lines associated with the National Synchrotron Light Source (NSLS) at the Brookhaven National Laboratory. Typical experiments on these beam lines involve a senior investigator and a very small number of associates—typically, a professor, a post-doc, and one or more graduate students. This sort of experiment has come to be known as *small science*. It is important to emphasize here that this *small science* is being carried out on a large machine and that the use of a large machine does not necessarily imply *big science*. I shall return to this question of *big* and *little science* when I discuss work in elementary particle physics.

The brightest X-ray source currently available is the Advanced Photons Source (APS) at the Argonne National Laboratory, and its use in Laue diffraction studies makes possible determination of the detailed structure of very complex molecules, such as the fragile histidine triad (FHIT) molecule shown in Figure 65.

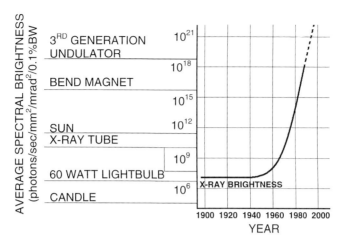

FIGURE 63   *The evolution of brightness of photon sources throughout the 20th century.*

FIGURE 64   *One sector of the beam lines from the National Synchrotron Light Source at the Brookhaven National Laboratory. (Courtesy of Martin Blume, Brookhaven National Laboratory.)*

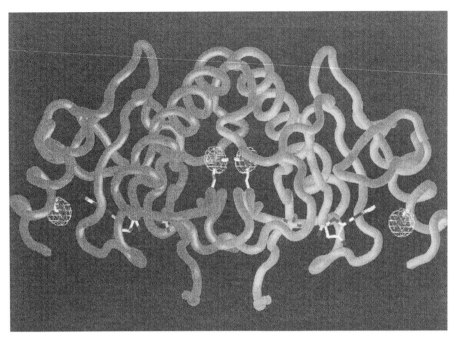

FIGURE 65 *The three-dimensional structure of the protein fragile histidine triad (FHIT) from x-ray diffraction at a resolution of 1.5 Å. The small white cages identify selenium atoms. (Courtesy of C.D. Lima and W.A. Hendrickson, Columbia University.)*

Improved Resolution and Higher Precision

One of the things that has characterized the last half of the 20th century has been the synergism between science and technology and the rapidly increasing precision arising from new instrumentation which evolved from earlier scientific discovery. Figure 66 illustrates the molecular spectroscopy of sulfur hexafluoride. Here, the resolution has increased by a factor of more than a million within the past 15 years. Indeed, the resolution shown in the bottom panel of this figure is limited simply by the Heisenberg Uncertainty Principle and by the finite time during which the molecules were under observation.

Figure 67, showing the specific heat of liquid ^{4}He in the vicinity of the

FIGURE 66 *Molecular spectroscopy (as in the case of sulfur hexafluoride, shown here) has increased in resolution by a factor of more than a million within the past two decades. Shown here are spectra of sulfur hexafluoride with progressive increases in resolution from top to bottom by factors of 100, 20 and 2000 respectively. The resolution in the lowest panel is limited by the Heisenberg uncertainty principle and by the finite time during which the molecules were under observation. (Courtesy of Canadian National Research Council.)*

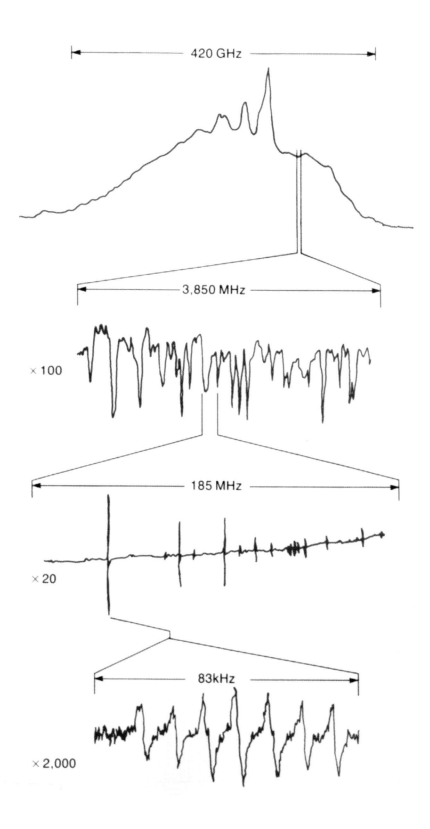

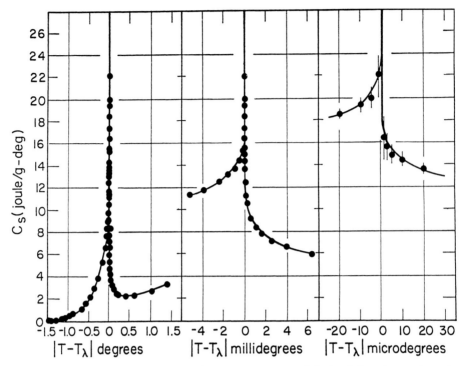

FIGURE 67 *The specific heat of liquid ^4He in the vicinity of the lambda point—where the transition from the normal (high temperature) to the superfluid (low temperature) phase occurs. It bears noting that the abcissae in the three sections of this figure are in degrees, millidegrees, and microdegrees, respectively, illustrating the detailed temperature dependent structure in the neighborhood of this second-order phase transition and the improvement in resolution from 1911 to about 1980. Currently, measurements can be made in nanodegrees. The solid lines are empirical fits to the experimental data points. (Courtesy of N.J. Buckingham and W.M. Fairbank.)*

lambda point, shows the dramatic progress made during the century. Note that the abscissa in the three panels changes by a factor of a million overall.

This increased instrumental resolution has made possible the measurement of physical constants with ever-increasing precision. Figure 68 lists the current 1999 best values for a number of these physical constants, and it bears noting that the uncertainty is typically a few parts in a million, in one hundred million, or in a billion except in the case of the Rydberg constant where the uncertainty is roughly one part in a million million.

Trapping and Cooling of Elementary Particles and Atoms

One of the other important developments in recent years has been the various kinds of devices wherein elementary particles and atoms or molecules can be trapped for indefinite periods and both cooled and probed by appropriate laser

SELECTED PHYSICAL CONSTANTS

Quantity	Symbol	Numerical value	Unit	Relative std. uncert. U_r
speed of light in vacuum	c, c_0	299792458	m s^{-1}	(exact)
magnetic constant	μ_0	$4\pi \times 10^{-7}$	N A^{-2}	
		$= 12.566370614... \times 10^{-7}$	N A^{-2}	(exact)
electric constant $1/\mu_0 c^2$	ε_0	$8.854187817... \times 10^{-12}$	F m^{-1}	(exact)
Newtonian constant of gravitation	G	$6.673(10) \times 10^{-11}$	m^3 kg^{-1} s^{-2}	1.5×10^{-8}
Planck constant	h	$6.62606881(51) \times 10^{-34}$	J s	7.8×10^{-8}
$h/2\pi$	$\hbar$	$1.054571605(82) \times 10^{-34}$	J s	7.8×10^{-8}
elementary charge	e	$1.602176469(62) \times 10^{-19}$	C	3.9×10^{-8}
magnetic flux quantum $h/2e$	Φ_0	$2.067833644(80) \times 10^{-15}$	Wb	3.9×10^{-8}
conductance quantum $2e^2/h$	G_0	$7.748091698(28) \times 10^{-5}$	S	3.6×10^{-9}
electron mass	m_e	$9.10938195(71) \times 10^{-31}$	kg	7.8×10^{-8}
proton mass	m_p	$1.67262160(13) \times 10^{-27}$	kg	7.8×10^{-8}
proton-electron mass ratio	m_p/m_e	1836.1526674(39)		2.1×10^{-9}
fine-structure constant $e^2/4\pi\varepsilon_0\hbar c$	α	$7.297352535(26) \times 10^{-8}$		3.6×10^{-9}
inverse fine-structure constant	α^{-1}	137.03599972(50)		3.6×10^{-9}
Rydberg constant $\alpha^2 m_e c/2h$	R_∞	10973731.568548(83)	m^{-1}	7.6×10^{-12}
Avogadro constant	N_A, L	$6.02214194(47) \times 10^{23}$	mol^{-1}	7.8×10^{-8}
Faraday constant $N_A e$	F	96485.3411(38)	C mol^{-1}	4.0×10^{-8}
molar gas constant	R	8.314473(14)	J mol^{-1} K^{-1}	1.7×10^{-6}
Boltzmann constant R/N_A	k	$1.3806504(23) \times 10^{-23}$	J K^{-1}	1.7×10^{-6}
Stefan-Boltzmann constant $(\pi^2/60)k^4/\hbar^3 c^2$	σ	$5.670400(38) \times 10^{-8}$	W m^{-2} K^{-4}	6.7×10^{-6}

Peter Mohr
Barry Taylor
NIST Physics
1999

FIGURE 68

irradiation while they are trapped. Figure 69 is a general schematic of the idea, but it should be emphasized that there are a wide variety of traps: the Paul trap and the Penning trap are similar from a geometrical point of view and use quadrupole magnetic fields. The Paul trap uses an intermediate frequency ac voltage between a hyperbolic ring and the hyperbolic end electrodes, while the Penning trap has a repulsive dc voltage on the end electrodes and a strong axial magnetic field. Hans Dehmelt at the University of Washington in Seattle, using a modified Penning trap, succeeded in trapping and holding a single positron for some three months (more than a hundred trillion times a positron's normal lifetime). By working with this single positron, he and his collaborators were able to measure the positron's magnetism and size to a precision one thousand times better than was achieved in any prior study. In fact, the Dehmelt group got to know this positron so well that they christened her "Priscilla!"

While visiting Dehmelt's laboratory a few years ago, I had an experience that I had been absolutely convinced would never be possible for any physicist. I was able to see, with my naked eyes, a single barium atom. In this case, Dehmelt had trapped a single barium atom and cooled it, using a laser cooling process

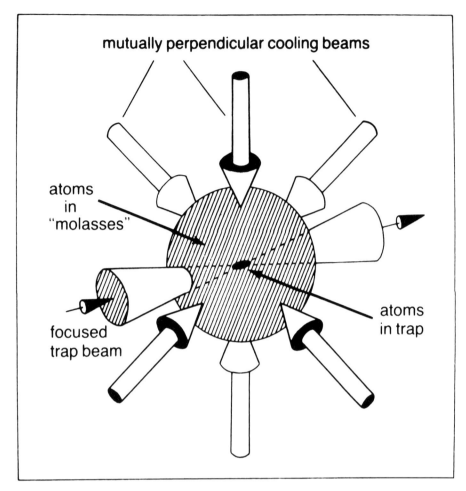

FIGURE 69 *A schematic illustration of the arrangement used to trap and cool elementary particles and atoms. (Courtesy of Hans Dehmelt, University of Washington.)*

he had invented, until the barium atom remained at rest in the center of a relatively large vacuum chamber. By optically pumping this atom (called Astrid) with an infrared laser to an excited state that emitted blue photons optically, it was possible to see the single atom through a window in the trap radiating these blue photons like a tiny blue star. Figure 70 is a photograph of Astrid radiating, and if you look very closely in the vicinity of the tip of the arrow, you too can see this single barium atom. For this and related work, Dehmelt shared the 1989 Nobel Prize with Wolfgang Paul and Norman Ramsey.

Bose–Einstein Condensates

In the early 1930s, Satyendra Bose and Albert Einstein predicted that a gas of noninteracting integer spin particles would condense into a macroscopic quantum state when cooled below a critical temperature. This so-called Bose–

Einstein (B-E) condensate has, of course, been seen in superfluid ^{4}He and in superconductors, but not, until very recently, in anything like a near-ideal gas. Over the past 15 years or so, with the availability of traps and laser cooling mechanisms, a number of groups have attempted finally to produce this Bose–Einstein condensation, but many of them used relatively light atoms. The first successful demonstration of the desired condensation was achieved by Eric Cornell and Karl Wieman, of the Joint Institute for Laboratory Astrophysics in Boulder, Colorado, together with Michael Anderson of the University of Colorado using ^{87}Rb atoms. The condensation occurred at a temperature of about 170 nanoK and contained something like 2000 atoms in the B-E single quantum state. Figure 71 shows the results obtained in these pioneering Colorado experiments.

Having B-E condensations (now of many atomic species), many groups are using the condensate as a physical probe. Many of these groups allow the condensate to fall under gravitational forces, and Figure 72 shows an experimental arrangement used by Michael A. Kasevich at Yale. Using such condensate arrays, he has developed a gyroscope with an accuracy better than 10^{-7} degrees per hour, which is at least a factor of hundred better than anything previously available, and a gravity gravimeter with a sensitivity between 30 and 40 times

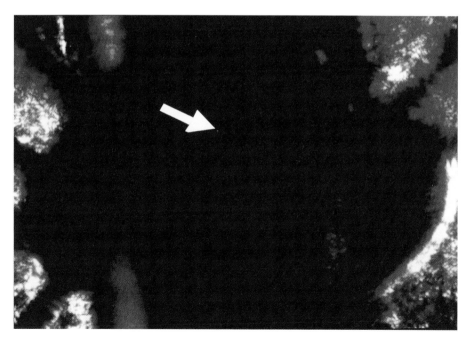

FIGURE 70   *A photograph of a single barium atom radiating blue light from an excited state into which it had been pumped by infrared laser photons. The barium atom was dubbed "Astrid" by Hans Dehmelt (1989) and his group at the University of Washington, and the author can vouch, from personal experience, that the light from this single barium atom was quite visible to the naked eye. (Courtesy of Hans Dehmelt, University of Washington)*

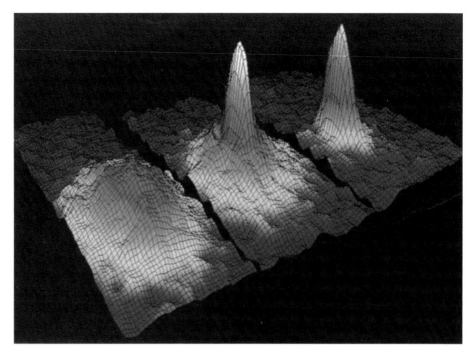

FIGURE 71   *An illustration of Bose–Einstein condensation in the velocity distribution of atoms in an evaporating cooling cloud of ^{87}Rb atoms. Before condensation begins (left), the distribution is isotropic, as expected for a gas in thermal equilibrium; the condensate appears (middle) as a fraction of the atoms that have velocity close to zero. The distribution is elliptical, as would be expected if all the condensed atoms are in the ground state of the elliptical potential. Continued evaporation leads to an almost pure condensation of about 2000 rubidium atoms (right). Each image in this photograph is 200 by 500 microns and was derived from the shadow of the atom cloud after 60 milliseconds of free expansion. (Courtesy of Michael Matthews, JILA.)*

better than anything previously available. We are only beginning to appreciate all that can be done with these B-E condensates now that we have them.

Perhaps the most surprising and new phenomenon using B-E condensates was that of Dr. Lene Hau (Figure 73) of the Rowland Institute for Science in Cambridge, Massachusetts, using a laser-dressed B-E condensate at 50 billionths of a degree Kelvin (one of the lowest temperatures ever obtained in the laboratory). Hau and her associates found that the refractive index of the condensate was about 10^{11} times greater than that of optical glass and, in consequence, a laser light beam tuned to one of the sodium resonance lines transversed the condensate not at the normal speed of light (see Figure 68) but at a rather remarkable 120 feet per hour. Dr. Hau is confident that she can bring this down to something like 12 feet per hour or less within the coming year. This work was simultaneously published in *Nature* and on the front page of *The New York Times* in March of 1999. It is too early yet to have any firm idea of how this new phenomenon will be used in physics or in technology, but already groups around the world are hot on the trail of such applications, whatever they may be.

Frontiers of Nuclear Structure Physics

Now we turn to nuclear physics. Figure 74 is a map of the nuclear domain originally drawn by Professor G.N. Flerov of the Soviet Union—the individual who had the distinction of having convinced Stalin to begin work on a Soviet atomic weapon and who was, for many years, the Director of the multi-nation Joint Institute for Nuclear Research at Dubna. As is clear from this map, the stable nuclei in nature form only a very small fraction of those radioactive isotopes that can be produced by a nuclear reaction induced by accelerated particles. The worldwide development of accelerated beams of short-lived radioactive nuclei themselves greatly extends the range of new species that can be produced and studied. There really are two current frontiers in nuclear structure physics. The first recognizes that what is shown in this figure is the map of the nuclear ground state and attention is focused on what happens as one moves up away from this ground state into the nuclear continuum. Second, there has always been a question of whether there were islands of stability beyond plutonium, the heaviest nucleus found naturally on the earth. Early shell models predicted that elements 114 and 126 might have sufficiently long lifetimes to allow their study using standard chemical technology.

Mark Kasevich 2000
Yale University
mark.kasevich@yale.edu

Experiment: Atomic tunnel arrays

Confining potential: Bose condensed atoms are loaded into a vertical lattice potential and subsequently tunnel to the continuum. Waves interfere to form pulses in region A.

Array Output:

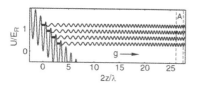

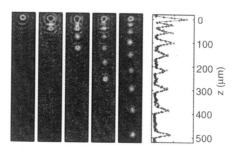

Measured pulse period of ~1.1 msec is in excellent agreement with calculated $\omega_j = mg\Delta z/\hbar$

FIGURE 72

FIGURE 73   *Dr. Lene Hau of the Rowland Institute for Science in Cambridge, Massachusetts with some of the equipment that she has used to demonstrate that the velocity of light in a Bose–Einstein condensation of sodium atoms at 50 millionth of a degree absolute is 120 feet per hour. The refractive index of the sodium condensate is a factor of 10^{11} times greater than that of optical glass. (Courtesy of Lene Hau, Rowland Institute for Science.)*

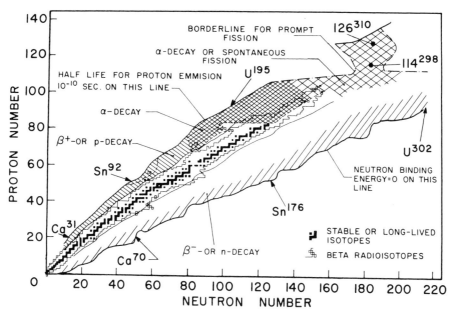

FIGURE 74   *A map of the nuclear ground states prepared by Professor G.N. Flerov of Dubna in the 1970s and modified by the author. It shows both the neutron and proton drip lines, which marked the boundaries of the region where radioactive isotopes have a finite lifetime, and demonstrates the wide range of isotopes that are theoretically possible for almost any element. In the case of uranium, these can range from ^{195}U to ^{302}U. Also shown here are the assumed doubly magic nuclei $^{298}114$, and $^{310}126$. But these numbers are based on relatively simple shell model calculations, while the more sophisticated ones predict that although the neutron magic number remains at 178 the proton magic number moves up to 120 or 126 and is not at 114. Current frontiers in nuclear structure involve exploring this map, particularly with the use of radioactive projectiles now being accelerated in new facilities in Japan, Canada, the U.S., and Europe; moving up away from this ground state map toward the nuclear continuum; and attempting to extend this map to even heavier species. (Courtesy G.N. Flerov, Dubna.)*

Three laboratories, those at Berkeley, at GSI in Germany, and at Dubna, have devoted enormous effort to searching for even heavier transuranic species. Figure 75 illustrates the technique used to make identifications when only a single atom of the desired nucleus is produced. In this work, carried out by Peter Armbruster and his colleagues at GSI, the doubly magic nucleus ^{208}Pb was bombarded with a beam of ^{70}Zn and the isotope of mass 277 of element of 112 was produced and subsequently decayed through a six-fold cascade of alpha particles eventually reaching ^{253}Fm. Because each of these steps had been carefully studied in the past and both the energy of the alpha particle and the corresponding lifetimes for its appearance have been determined with precision, this chain of alpha particle decays provides a unique signature for element 112 despite the fact that only a single nucleus had been produced.

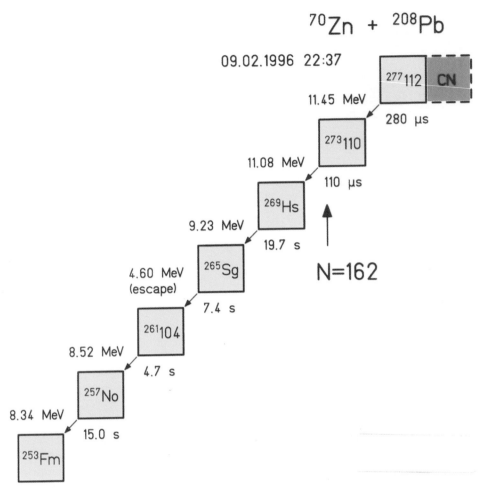

FIGURE 75   *The decay schematics of the isotope $^{277}112$ as determined by Peter Arm-bruster and his collaborators at GSI, Darmstadt, Germany. The ^{208}Pb target was bom-barded by a beam of ^{70}Zn nuclei forming a single nucleus of element 112, and after loss of a single neutron this nucleus decayed in the illustrated six-fold alpha particle cascade ending in ^{253}Fm. This examination of a decay alpha particle cascade has become a stan-dard method for identifying new elements as they are produced in heavy ion collisions. (Courtesy of Peter Armbruster, GSI.)*

Early in 1999, the group at Dubna, led by Yuri Oganessian, announced that it had produced element 114 by bombarding ^{244}Pu with ^{48}Ca; they identified the announced element 114 by studying a three-stage alpha particle cascade in its decay. What was remarkable was that its half-life was found to be 30 min-utes, vastly longer than that of any of the other transuranics in the region. This might be expected if 114 were, in fact, a magic number for protons; but what causes some uncertainty is the fact that only a single, relatively simple, shell model predicts that to be the case; more sophisticated versions of the shell model predict that 120 or 126 are more probably the proton magic number. Both

Berkeley and GSI groups are now at work attempting to duplicate the Russian measurement, and the research for ever-heavier transuranics and possible islands of stability continues.

One of the very interesting new results in moving away from the nuclear ground state was the discovery that nuclei having prolate ground states, at higher excitation typically had both superdeformed and, in some cases, even hyperdeformed classes of states in potential minima having ever stronger deformation. The first experimental evidence for superdeformation in dysposium nuclei is shown in Figure 76. The announcement was made by Peter Twin and his collaborators from the Daresbury Laboratory in England that the superdeformed nucleus created in heavy ion collisions, with an angular momentum of 60 units, decayed by emitting successive quadrupole photons in a cascade, here identified down to a spin of 26 units. The effective moment of inertia and thus the deformation of the nucleus in this superdeformed state can be obtained from the energy and energy spacing of these gamma ray lines while the quadrupole moment

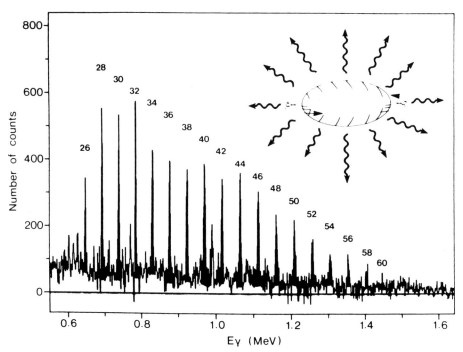

FIGURE 76   *The gamma ray spectrum from superdeformed dysprosium nuclei produced in heavy ion reactions and initially formed here with an angular momentum of 60 units; the decay cascade in the superdeformed state involves emission of quadrupole photons. The spacing of the gamma ray lines and their energies gives the moment of inertia of the superdeformed nucleus while their lifetimes gives its quadrupole moment. This figure shows the original measurements of Peter Twin and his associates at the Daresbury Laboratory in England. Such studies have been among the most active in nuclear physics over the past two decades. (Courtesy of Peter Twin, Daresbury.)*

can be obtained from their lifetimes. The study of such gamma rays has become one of the most active areas worldwide in nuclear structure physics.

Figure 77 illustrates an interesting aspect of these studies reported by T.L. Khoo and his collaborators from the Argonne National Laboratory. This schematic figure shows the energy levels as a function of deformation showing the normal ground state minimum and the superdeformed minimum. The original heavy ion collision (a) populates chaotic states in the normal nuclear continuum. These then decay via gamma ray emission (b) into highly ordered low-lying states in the superdeformed family of states. They, in turn, decay (c) into still-chaotic states in the normal minimum, and finally these decay into lower-lying states in that family that are highly ordered. This cascade from chaotic to

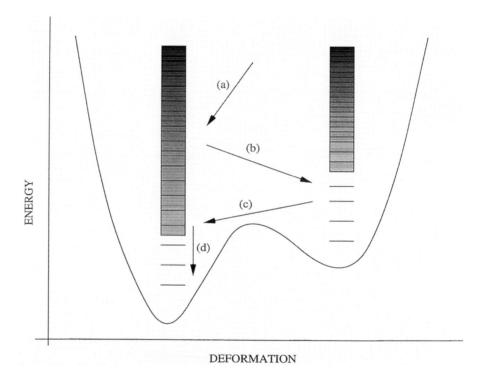

DEFORMATION

FIGURE 77   *A highly schematic representation of experiments carried out by Teng-Lek Khoo and his associates at the Argonne National Laboratory on the gamma ray decay of superdeformed nuclei. The original heavy ion collision populates states of chaotic character in the continuum of the normal nucleus, i.e. in the least deformed of the minima in the potential surface. Some of these states decay by gamma ray emission to low-lying ordered states in the superdeformed family which in turn decay back to chaotic states in the normal nucleus and finally decay to highly ordered low-lying states in that nucleus. The important point about these measurements is that they have demonstrated that chaotic and highly ordered states of different deformation occur at precisely the same energy in these nuclei and thus give an excellent opportunity to study the characteristics of quantum chaos as it appears in the nuclear system. (Courtesy of T-L Khoo, Argonne National Laboratory.)*

FIGURE 78   *Niels Bohr (1922) at the blackboard discussing elements of nuclear structure with his son Aage Bohr (1975). Aage Bohr (1975), working with Ben Mottelson (1975), was one of the primary figures involved in developing the collective model of nuclear structure. (Courtesy of AIP Emilio Segré Visual Archives, Margrethe Bohr Collection.)*

orderly to chaotic to orderly has been studied, and the point of greatest interest is that chaotic and orderly states in the one nucleus can coexist at the same energy, allowing the possibility of detailed study of the nature of these states, of their interaction, and of the quantum chaotic process itself.

Figure 78, taken in the 1960s, shows Niels Bohr explaining a fine point in nuclear physics to his son, Aage—emphasizing the fact that Niels Bohr maintained his interest in nuclear physics despite being one of the more active international proponents of world peace for many years. The late Sven-Gosta Nilsson told me, many years ago, of an interesting experience he had as one of Aage Bohr's graduate students. Toward the end of his dissertation work, Nilsson had encountered a subtle theoretical problem. Even though it was a Sunday morning, he called Aage Bohr's home in hope of getting some help. The phone was answered, and in his urgent need of help Nilsson spelled out his problem and received a very clear, concise, and complete answer to it. As the answer proceeded, however, Nilsson began to recognize that Aage's voice seemed a little different; finally, when the answer was complete, Nilsson asked "This is Aage Bohr, is it not?"; to which he received the answer, "No—this is Niels Bohr." He had remained fully in touch with even the most subtle problems in nuclear structure.

The Phase Diagram for Nuclear Matter

In moving to ever-higher energies in the nucleus, we must necessarily consider the phase diagram for nuclear matter shown, in one particular form, in Figure 79, which plots the temperature versus the baryon density. We know that this diagram was traversed in the early universe when the cooling primordial plasma condensed to hadrons at a temperature of about 200 MeV and a very low density, and we also know that in supernovae the diagram is traversed at relatively low temperature but at very high density. The fact that the phase change from hadron matter to a plasma of deconfined quarks and gluons is not sharp simply reflects the finite size of the nucleus. The liquid gas phase at densities less

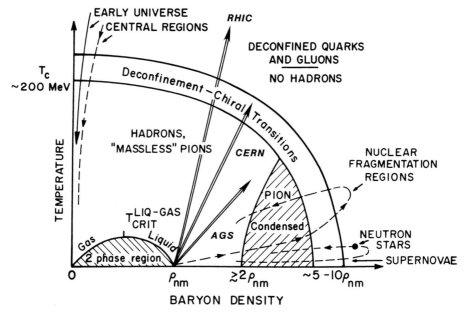

FIGURE 79   *A phase plot for nuclear matter showing temperature in MeV versus density in units of the normal nuclear density. The phase transition from hadronic to quark matter is not a sharp one because of the finite size of the nuclear system. Neither the postulated liquid-gas at low density nor the pion condensed states at higher density have ever been observed. As indicated, the Alternating Gradient Synchrotron (AGS) used in its heavy ion mode at the Brookhaven National Laboratories cannot reach sufficient energy to bring about the transition while CERN, in its present state, barely reaches the transition. The Relativistic Heavy Ion Collider (RHIC) just coming into operation at Brookhaven, on the other hand, is expected to probe the quark-gluon plasma and, when completed, the Large Hadron Collider (LHC) at CERN will have an energy 40 times that of RHIC in its collision mode and thus will explore the quark-gluon plasma region in detail.*

than that of normal nuclear matter has never been detected nor, indeed, has the predicted pion condensate between roughly two and five times normal nuclear density. As shown here, the alternating gradient synchrotron (AGS) at Brookhaven does not accelerate heavy ions to sufficient energy to create the quark–gluon plasma, and CERN in its present condition barely reaches the transition from hadronic to quark–gluon plasma.

The relativistic heavy ion collider (RHIC) at Brookhaven, which began operation in 1999, has already shown evidence of taking us into the quark–gluon region, and the large hadron collider (LHC) now under construction at CERN, with an energy 40 times that of RHIC, will certainly take us into entirely new regions. The need for these high energies and for the heaviest possible targets and projectiles is illustrated in Figure 80, is a highly artistic view of a heavy ion collision at relativistic energies. As shown in the top panel, the projectile is relativistically contracted into a disc approaching the target nucleus from the left. As it traverses the target, the nuclear matter in both is raised to very high temperature; as the projectile passes through the target, what is left are two regions of high baryonic density—the remnants of the target and the projectile—while between the two is the so-called *fire tube* with low (hopefully zero) baryonic density, where we would anticipate finding the quark–gluon plasma. The higher the energy, the purer this quark–gluon plasma will be and the less contaminated by baryonic fragments.

Figure 81 is a photograph of the Brookhaven site showing the tandem accelerators at the lower right, where the initial acceleration of nuclear species up to gold is accomplished. The accelerated beam is then transferred through a transfer line to the AGS and to the booster, which raises the energy before final acceleration in AGS and injection into the large RHIC rings shown at the top of this photograph. One of the major challenges, once the RHIC accelerator becomes operational, is that of finding whether, in fact, strange matter involving strange as well as up and down quarks exists in stable form as predicted on the basis of current understanding (see Figure 92). Such strange matter could, in principle, occur in units with masses up to 10^{57} atomic units! Figure 82 shows the STAR detector that is currently being installed in one of the six intersection regions where the countercirculating heavy-ion beams are brought into collision in the RHIC system. John Harris of Yale is the spokesperson for this detector system, and a few of the individuals who have been heavily involved in its construction and installation are shown in this photograph.

Atomic and Nuclear Technology in Biology and Medicine

Atomic and nuclear technology have found wide application in biology and in medicine, and the interconnections are growing on almost a daily basis. In the next few figures I show only a few representative illustrations.

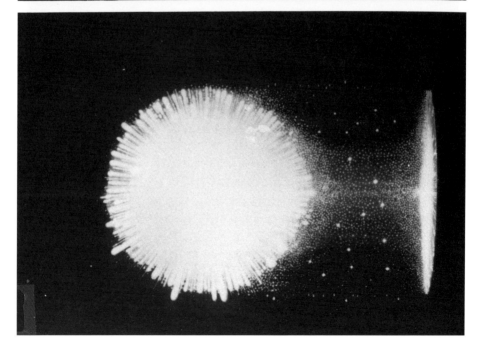

FIGURE 80   *A highly artistic view of a relativistic collision of two heavy nuclei. In the upper panel, the projectile coming in from the left has been relativistically foreshortened to appear as a disc. In the middle panel, as it traverses the target nucleus, both have their temperatures raised to very high values and after the projectile has passed completely through the target, there remain two highly excited, high-baryon-density objects separated by the so-called fire tube of very low baryonic content and it is in this fire tube that one expects to be able to examine the quark gluon plasma and how it condenses from its initial high temperatures. Since it is already known that quarks of other than the first family are produced in these relativistic heavy ion collisions there is the hope of producing strange matter involving the higher quark families. (Courtesy of Hans Gutbrot, Lawrence Berkeley Laboratory.)*

FIGURE 81   *A photographic overview of the Brookhaven RHIC site. At the lower right is the double tandem accelerator which produces beams of all stable nuclear species up to and including gold for transfer through the indicated beam line to the alternating gradient (AGS) site then to the new booster that has been developed to increase the output energy of the AGS for injection into the two RHIC rings in the tunnel shown at the top of this figure. There are six regions in the ring where the beams in the countercirculating storage rings are brought into collision and where massive detection equipment is being established to study the products of these collisions. (Courtesy of Martin Blume, Brookhaven National Laboratory.)*

FIGURE 82   *A view of the STAR detector during its installation on the RHIC ring. Typical of such detectors, it contains high resolution silicon detectors surrounding the immediate collision region, then a large time-projection chamber and finally an array of scintillation and ionization detectors all enclosed in a very large magnet. This entire detector can be rolled on its own rails away from the collision point for adjustment and modification. (Courtesy of John Harris, Yale University.)*

Using technology originally developed for electron accelerators, it has been possible to produce electron microscopes that give an entirely new visceral insight into biological systems. Figure 83, showing a human cancer cell reaching out to attack a substrate of normal cells, is an example of such insight.

One of the most widespread of human scourges is blockage or other damage to cardiac arteries; and although standard X-ray technology, with the injection of iodine or other heavy contrast medium, has made possible literally millions of angioplasty procedures and the installation of metal stents to prevent reclosure of constricted arterial passages, there is a continuing demand for better visualization of the cardiac circulation. Figure 84 shows an example of the detail that can be obtained by using so-called digital subtraction angiography, which greatly reduces background and gives a very precise image of the cardiac circulation.

Figure 85 is an example of the use of both normal nuclear magnetic resonance imaging (MRI) and functional nuclear magnetic resonance imaging (fNMRI) in a study of autism in the human brain. As is clearly evident from a comparison of the activity shown here in color in a normal brain and an autistic one when presented with a facial recognition task, it is clear that the autistic brain shows characteristically less activity, consistent with an autistic patient's inability to recognize other persons. This combination of structural and functional information has given medical science an entirely new insight into the human central nervous system, and senior biologists—including Philip Handler, former President of the National Academy of Sciences, and Harold Varmus, current Director of the National Institutes of Health—have repeatedly stated that the use of modern physics, technology, and understanding has moved our understanding of the human central nervous system and brain forward more in the past five years than in all of prior history.

Some measure of the importance of these developments comes from the fact that one in four Americans at some point in life will suffer a brain-related

FIGURE 83   *A three-dimensional view of a human cancer cell, magnified 3000 times, showing how these cells reach out to engulf neighboring normal cells. (Courtesy of E.J. Ambrose.)*

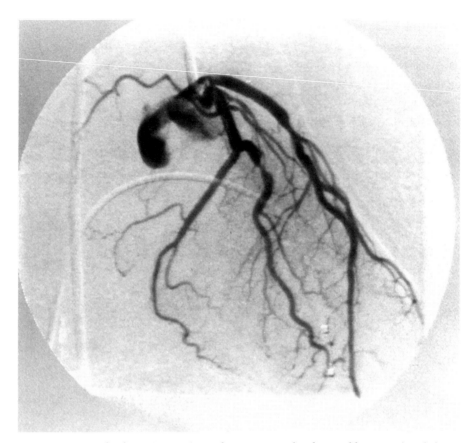

FIGURE 84   *Digital subtraction angiography. An example of ensemble processing, it is used to visualize the cardiac circulatory system, which is problematical for diagnosticians because it has almost the same X-ray absorption properties as do the tissues surrounding it. To visualize the circulatory system, here involving the heart itself, an X-ray image is made by directing an X-ray beam through the patient to a high resolution video camera. This produces a background image. A contrast dye (usually containing iodine) is then injected into the patient's circulatory system—usually through a catheter into the heart itself—and another X-ray image is made. The two images are then aligned in a computer and the second image subtracted digitally from the first. The result, as shown here, is a clear image of the circulatory system without the background tissues. A very similar subtractive approach is taken in angiography using very short synchrotron x-ray pulses where the x-ray energy is adjusted just above and then just below the K absorption edge for the iodine dye and the two resulting images are timed to match the same point in the heartbeat and are then subtracted to again yield a very clear background free image of the coronary circulation. (Courtesy of Siemens Medical Systems, Inc.)*

disorder. In the United States, such brain-related disorders cause more people to be hospitalized than any other major disease group, including cardiovascular disease or cancer, and treatment, rehabilitation, and related consequences of brain-related diseases cost Americans an estimated $400 billion each year.

Figure 86 is yet another specific illustration of the combined use of positron emission tomography and nuclear magnetic resonance in the treatment of epileptic patients. By subtracting PET/SPECT images superimposed on MRI images of a patient's brain when in the normal state and when in an epileptic seizure, the surgeon has a precise indication of the malfunctioning region of the brain, which he or she can then remove surgically.

Figure 87 shows one of the most important developments of the entire century—ranking with relativity and quantum mechanics. Based to a large extent on the X-ray diffraction studies of the DNA molecule that had been carried out by Rosalind Franklin and Maurice Wilkins, James Watson and Francis Crick deduced the double helix structure of the DNA molecule—and for this they received the Noble Prize for Medicine and Physiology, with Wilkins, in 1962.

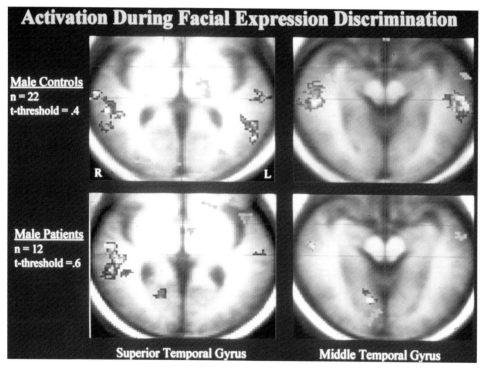

FIGURE 85 *Comparison of a normal and an autistic human brain when faced with a facial recognition task. The colored overlays are those obtained in functional MRI and the background structural information comes from normal MRI. It is clear that the activity in some parts of the autistic brain is almost totally missing as compared to the normal situation. (Courtesy of James Duncan, Yale University.)*

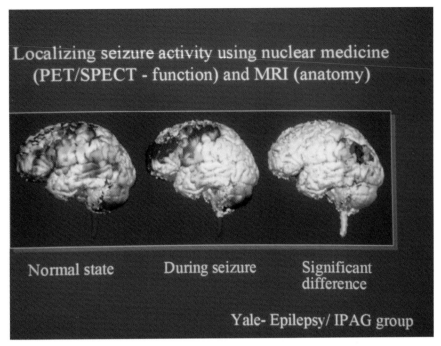

FIGURE 86   *An illustration of the use of positron emission tomography (PET) together with MRI to localize the areas involved in epileptic seizures so that they can be surgically removed. The figure on the left shows the patient's brain in a normal state with the PET information superimposed on that from the MRI. This is repeated during a seizure, and the two images are then subtracted to yield that on the right which identifies, in color, the specific brain areas involved in the seizure and thus the target for surgical intervention. (Courtesy James Duncan, Yale University.)*

The Merging of Elementary Particle Physics and Cosmology

Now we turn to two fields that are slowly coming together to address some of the most fundamental questions in physics—the fields of elementary particle physics and cosmology. They are coming together because, with ever more powerful accelerators, it becomes possible to recreate—if only for tiny fractions of a second—the conditions that were present within the first moments of the existence of our universe. Figure 88 is an interior view of the Fermi National Accelerator—currently the highest energy accelerator in the world—showing both the upper room-temperature ring and the lower superconducting ring and giving some idea of scale. Figure 89 shows that over time, the available beam energy from accelerators has been increasing in almost a perfect exponential, although the technology required to achieve these energies has been evolving and changing steadily. Figure 90 is an aerial view of the French–Swiss border region showing the two main CERN accelerator rings superposed. Figure 91 shows the detector system used at CERN by Carlo Rubbia and his associates in their original detection of the Z^0 boson for which he and Simon Van der Meer received the 1984 Nobel Prize in Physics.

The Standard Model

Beginning with Murray Gell-Mann and George Zweig at Cal Tech, who proposed the quark model, the so-called "standard model" has emerged from measurements made worldwide using high-energy accelerators. This standard

FIGURE 87   *James Watson and Francis Crick with their first model of the DNA double helix that they constructed from the x-ray diffraction information obtained by Rosalind Franklin and Maurice Wilkins. This discovery has had a revolutionary impact on modern biology, and the sequencing and mapping of human DNA is already well underway. Watson and Crick did their work in 1952, published it in Nature in 1953 and received the Nobel Prize for Medicine and Physiology, together with Maurice Wilkins, in 1962. In mid-2000 a private sector group led by Craig Venter and a federally-supported one led by Francis Collins simultaneously published identical and complete maps of the human genome. There were two surprises; first that this mapping could be completed so soon and second that instead of the long anticipated 100,000 genes only 30,000 are present and many plants are known to have more than 25,000 genes. This demonstrates that the traditional view that each gene was responsible for one and only one protein is incorrect and leads to the whole new scientific field of protenomics. (Copyright A. Barrington Brown—Photo Researchers, Inc.)*

FIGURE 88   *A photograph showing the tunnel of the main accelerator at the Fermi National Accelerator Laboratory. The upper magnet ring, shown in red and blue, is that of the original 400 GeV proton syncrotron accelerator. The lower ring, shown in yellow and red, comprises the superconducting magnets which were installed much later than the original ring to provide the colliding beams, each at 400 GeV, required for the new Tevatron accelerator. The Fermi Laboratory accelerator ring is some 4 km in diameter, and currently the accelerator produces the world's highest collision energies. Its most recent major discovery was that of the top quark. (Courtesy Fermi National Laboratory.)*

model is based on the building blocks of nature shown in Figure 92 undergoing interactions described by electroweak theory and quantum chromodynamics. As indicated in the figure there are three families of quarks and leptons—with matter, as we know it in our normal everyday existence, being composed entirely of members of the first family, the up and down quarks together with the electron and the electron neutrino. The second and third families have been found only in accelerator experiments, and it is rather comforting to know that there *is* no fourth family. This follows from the width of the Z^0 boson and from the relative abundance of hydrogen, helium, and lithium in the universe. The very heavy top quark was discovered at the Fermi National Laboratory, and during the past five years there has been what appears to be relatively solid evidence for the existence of a small but nonzero mass for the electron neutrino. This evidence comes from Japanese observations on solar neutrinos reaching their detectors directly and then, 24 hours later, reaching them after penetrat-

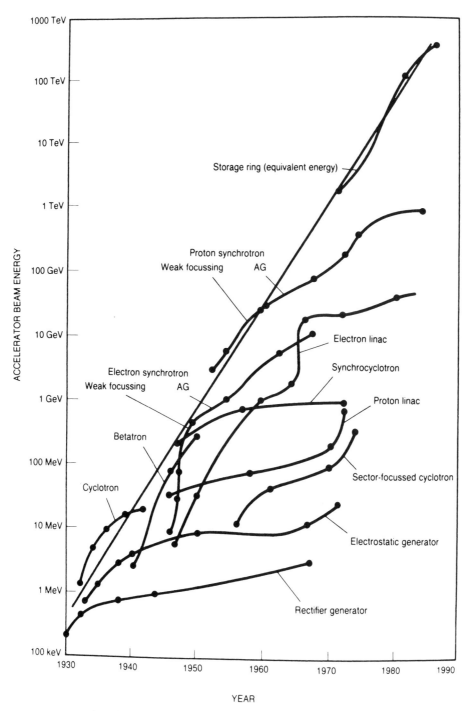

FIGURE 89　*A plot of the accelerator beam energies versus time for the period since 1920. It is striking that the output energy increases in almost perfectly exponential fashion, although the specific technology required to achieve this exponential increase changes as indicated—from the Cockcroft–Walton generator at the low end, to the storage ring colliders at the upper end.*

FIGURE 90 *Aerial view of CERN and the surrounding region. Three rings are visible. The smallest shows the underground position of the PS, the middle ring is the SPS (with a circumference of 7 km), the largest ring (27 km) is the LEP. Part of Lake Geneva appears in the background. (Courtesy of CERN.)*

ing the earth. If the neutrino has a nonzero mass, then we believe that there can be oscillations between the various neutrino families. These questions will hopefully be resolved when the Sudbury Neutrino Observatory in Canada becomes operational. This observatory is located under more than a mile of the nickel–iron remnant of a meteorite that struck the earth in prehistory and which has served as the richest source of nickel ore available worldwide. The heart of the detector comprises about 100,000 gallons of heavy water shielded with ordinary water in a large cavity carved out under the meteorite remains. Its advantages are that it can detect all neutrino families as well as neutrinos from the proton–proton cycle in the sun. It is currently in the final stages of construction.

The Building Blocks of Nature

In addition to the building blocks shown in Figure 92, there are, of course, the forces of nature, ranging from the strong nuclear force through the weak nuclear force to electromagnetism and, finally, gravitation. Figure 93 shows that current knowledge would suggest that were we able to reach energies of 10^{17} GeV—corresponding to an age of the universe within 10^{-38} seconds with a temperature of 10^{30} K, we can expect a grand unification of the electroweak and strong forces. Unfortunately, there is no hope whatever of our being able to reach energies of this magnitude on this planet.

As the most fundamental of the sciences, physics spans a rich spectrum of phenomena, and there has been a natural tendency to evolve subfields with their own vocabularies and their own techniques and technologies. Communication

FIGURE 91   *A view of part of the UA1 detection system used by Carlo Rubbia (1984) in the discovery of the Z^0 intermediate bosons at CERN. (Courtesy of CERN.)*

THE BUILDING BLOCKS OF NATURE

Quarks

	d	u	s	c	b	t
Charge	- 1 / 3	2 / 3	- 1 / 3	2 / 3	- 1 / 3	2 / 3
Mass (MeV)	4	7	150	1300	5500	170000
Spin	1 / 2	1 / 2	1 / 2	1 / 2	1 / 2	1 / 2
Strangeness	0	0	- 1	0	0	0
Charm	0	0	0	1	0	0
Beauty	0	0	0	0	- 1	0
Truth	0	0	0	0	0	1

Leptons

	$V\varepsilon$	e	$V\mu$	μ	$V\tau$	τ
Spin	1 / 2	1 / 2	1 / 2	1 / 2	1 / 2	1 / 2
Charge	0	- 1	0	- 1	0	- 1
Mass (MeV)	$<10^{-8}$	0.5	<0.3	105.6	<35	1784.1

FIGURE 92 *The building blocks of matter.*

between and among the subfields has always been less than it could and should be. Figure 94 simply reminds us that there is still a remarkable unity underlying all of physics. What I show here is simply a set of spectra from four different subfields of physics. The exiton spectrum of cadmium selenide represents condensed matter physics and that with the potassium atom, atomic physics. The lower left and right panels represent nuclear and elementary particle physics, respectively.

The most striking problem with the standard model is simply that it does not bring together gravity with the other forces of nature. This unification is one of the grand challenges in physics.

String Theory

In 1974, Joel Schert and John Schwarz proposed that, because string theory incorporated general relativity as well as gauge theory and was ultraviolet finite, it should be considered a candidate for a unified quantum theory of gravity and all other forces. This meant that the characteristic string lengths scale should be close to the Planck scale (10^{-32} cm) rather than the QCD scale (10^{-13} cm). Equivalently, the string mass scale should be about 10^{18} GeV, rather than 100 MeV. A huge amount of effort has been devoted to this whole question of whether the unification of gravity with the other natural forces can be handled in one of the variants of superstring theory. Within the theory—which is customarily cast in a ten-dimensional universe, six of which compactified in the

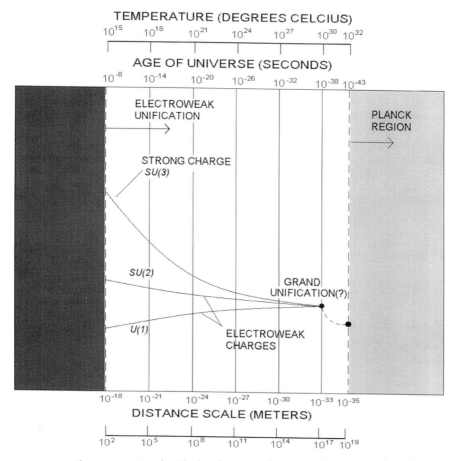

FIGURE 93   *Charges associated with the electro weak force and the strong force depend on the separation between the particles carrying the charges. Quantum theory leads to fluctuations in energy which are manifest as a sea of "virtual" particles throughout space. The virtual particles carry charges that can shield the separated particles, thereby causing the variation in the strength of the charges. The other three horizontal coordinates indicate equivalent ways of understanding variation. If one extrapolates from the measured values of the charges at an energy of 100 GeV, the three charges appear to be equal in strength at about 10^{17} GeV. The extrapolation suggests that a larger grand unified symmetry prevails above that energy in which the electroweak and strong forces become indistinguishable. The grand unified energy, however, may not be distinct from the energy at the Planck scale (10^{19} GeV), in which case the force of gravity cannot be neglected.*

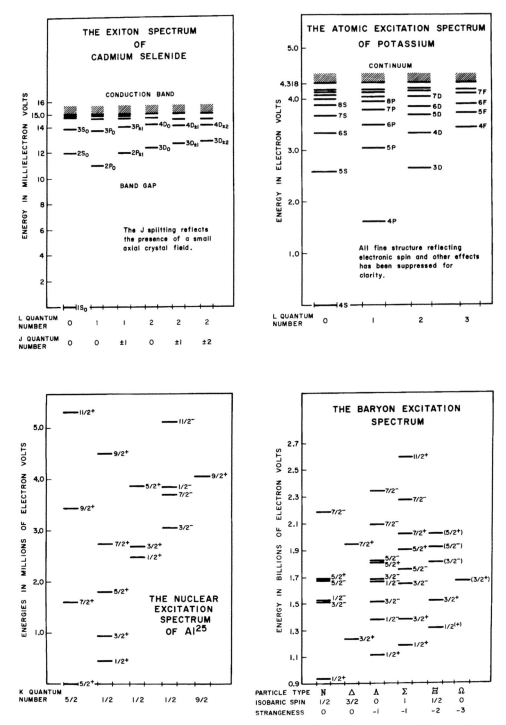

FIGURE 94 *Quantum excitation spectra for distinct physical systems. In each case, the spectra have been drawn to show typical classification of quantum states into families having common quantum numbers. Particularly striking is the similarity in the general appearance of these spectra despite a range of a million million (10^{12}) in the excitation energies involved.*

earliest instant of the universe to leave our familiar three spatial dimensions and one temporal one—the elementary particles are thought of, in theory, as one-dimensional curves (strings) rather than as points. Obviously, string theories can involve two quite distinct topologies: the strings can be open-ended or closed into loops. Currently, there are three known consistent types of string theories. Type I involves both opened and closed strings, while Type II and its variants and heterotic theories involve only closed loops. There is currently no way of choosing between them. More recently, in 1994, a new eleven-dimensional superstring theory (sometimes called membrane theory) was introduced.

The extra dimension means that the elementary particles are no longer considered as strings (one-dimensional) but rather as surfaces (two-dimensional), but in the same characteristic dimensions and masses. Obviously, the surfaces have many more normal modes than do the strings and thus can represent a much more complex array of elementary particles. Although this new membrane theory still remains essentially impregnable to experimental test, its practitioners have a level of optimism that has not been seen in this field for decades.

Figure 95 is a highly schematic representation of some string dynamics, but because I quite frankly do not fully understand all the latest details of string theory, let me simply leave it at that. We have not yet achieved a truly unified theory, but the experts appear optimistic.

Plasma Physics

Because more than 98% of the known universe is in a plasma state, I would be remiss if I did not at least refer briefly to the present situation in plasma physics. Much of the activity over the past 50 years has been driven by the hope of achieving nuclear fusion in either a magnetically confined plasma or an inertially confined one driven by either high-energy laser or high-energy particle beams symmetrically focused on a fuel target, which suffers resultant compression and heating to the point of achieving fusion.

In 1952, a group at Harwell in England announced that it had observed neutrons from a confined plasma, and the assumption was that fusion energy was just around the corner—energy too cheap to meter was sometimes discussed! It subsequently turned out that the neutrons had come from the DD reaction and deuterons accelerated by local irregularities in the magnetic confinement field. More realistic estimates suggested that perhaps in 20 years fusion energy would be available (the engineering problems of confining a plasma at a temperature of something approaching 100 million degrees and at a pressure in excess of 20 million pounds per square inch were better appreciated). Although there has been extensive work in many countries over the period since 1952, even the experts concede that we are still something like 20 to 25 years

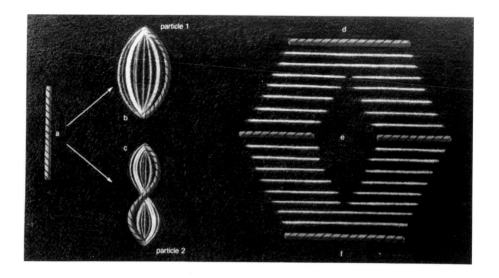

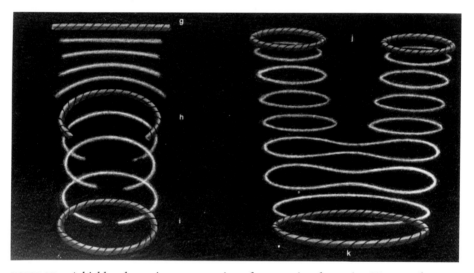

FIGURE 95 *A highly schematic representation of some string dynamics. Here an elementary string (a) can vibrate in different modes to represent two different particles (b) and (c); a single string (d) can divide (e) to represent a decay process or join (e) and (f) to represent a fusion process. An open string (g) can bend (h) and join ends to form a closed loop (i); two such loops can then join to represent a fusion process of (j) → (k) or separate, (k) → (j) to represent a decay. (Courtesy of F.E. Close, Rutherford Laboratory.)*

away from any hope of fusion energy because the problems have turned out to be more serious than we had ever anticipated.

Figure 96 shows an engineer doing repairs inside the toroidal plasma space of the General Atomics Tokomac magnetic-confinement system, while Figure 97 is a view of the main laser amplifier chains in the old NOVA laser facility at the Lawrence Livermore Laboratory. A much larger laser facility of the same type is now scheduled for construction at Livermore as the National Ignition Facility (NIF). Figure 98 is a streak camera photograph showing the compression and ignition of a glass mircoballoon target of deuterium and tritium driven by the OMEGA high-power laser at the University of Rochester. This installation for the next several years will be the world's most powerful—until the NIF becomes operational.

One of our most successful international cooperations in physics had been the ITER (International Experimental Thermonuclear Reactor) program equally shared by Russia, the European Community, Japan, and the United States. Unfortunately, budget constraints have forced the United States to cut back on its support of the ITER project in order to maintain the program of plasma physics studies in universities and national laboratories across the United States, and it is not yet clear whether this action has effectively terminated the ITER project or whether our former partners will continue without us as they had initially suggested might be the case.

This is a most unfortunate situation because it reinforces the widely held U.S. reputation as "an unreliable partner" in major international scientific and

FIGURE 96 *An engineer makes adjustments at the General Atomics DIII-D fusion machine inside the toroidal confinement space in which the interacting—and, hopefully, fusing—plasma is contained. (Courtesy of General Atomics.)*

FIGURE 97   *A view of some of the laser amplifier chains in the Lawrence Livermore National Laboratory NOVA laser system. This system has now been closed down in preparation for the construction of the National Ignition Facility. (Courtesy of the Lawrence Livermore National Laboratory and the U.S. Department of Energy.)*

technological cooperation; given the overall budgetary situation for U.S. plasma physics, there really is no other choice if we are to maintain the scientific research programs in plasma physics that will be essential to progress when the U.S. funding situation improves.

Plate Tectonics in Geophysics

In the class of major surprises and gross upheavals in the fundamental understanding within a field, I would be remiss not to mention the recognition of plate tectonics in geophysics. I remember vividly, in 1960, watching Tuzo Wilson of the University of Toronto laughed off a platform for suggesting that a Pacific tectonic plate moved relative to the planet and that its motion over a hot spot in the Pacific gave rise to the line of Emperor Seamounts and the Hawaiian Islands. This idea of continental drift was not new and had been suggested by Taylor in 1910 and by Wegener in 1912. In the absence of any recognized mechanism capable of driving the plate motion, however, these suggestions had never been considered seriously. This, in retrospect, is surprising, because in Figure 99 it is clear that even a superficial glance at the shape of the continents would suggest that sometime in the past they had been much closer together than they are today. In the last four decades, major progress has been made in identifying the plates and in recognizing the plate boundaries and their enormous impact on the surface structure of our planet.

The Structure of Supernovae

Throughout history, it has been recognized that in more or less random fashion, certain stars underwent cataclysmic explosions—supernovae—and for a short time emitted energy at truly astronomical rates. Until 1987, what was ac-

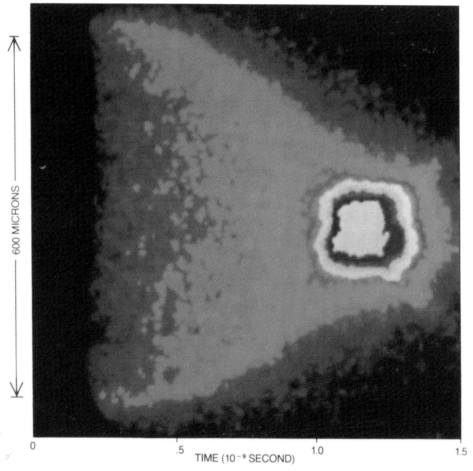

FIGURE 98   *A slit camera photograph showing the compression and ignition of a spherical fuel pellet during irradiation by the laser beams from the Omega Laser System at the University of Rochester. (Courtesy of Robert McRory, University of Rochester.)*

tually happening in one of these explosions was necessarily purely conjecture incorporating everything that we knew about the physics of the time. Figure 100 is a highly schematic illustration of what we believed had happened if a star ran out of hydrogen fuel and, in consequence, began to collapse under gravitational forces. What happened subsequently depends upon the initial mass of the collapsing star as was originally calculated by Chandrasekar. If the initial mass is less than 1.4 times that of the sun, electron degeneracy pressure stops the collapse, and if less than twice the mass of the sun, neutron degeneracy halts the collapse. If the initial mass is more than three times that of the sun, however, a black hole becomes inevitable. An enormous amount of effort has been devoted in the attempt to understand a stellar collapse in detail and in one of the most complete among such studies, Gerald Brown and Hans Bethe made detailed predictions for how the collapse would proceed and

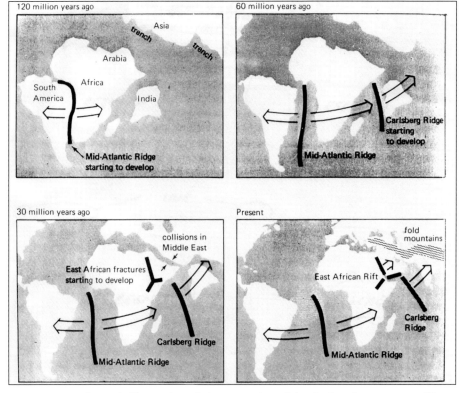

FIGURE 99   *A schematic illustration of plate tectonic activity during the past 120 million years. The motion of the tectonic plates has led to a number of striking features, such as the Mid-Atlantic Ridge, the Himalayan Mountains, the African Rift, and the Carlsburg Ridge. The processes that drive these plate motions are not yet fully understood, though convection in the mantle clearly is involved. (Courtesy of Science Year, 1968, Field Enterprises Educational Coporation.)*

Matter's Struggle for Space in the Crush of Collapse

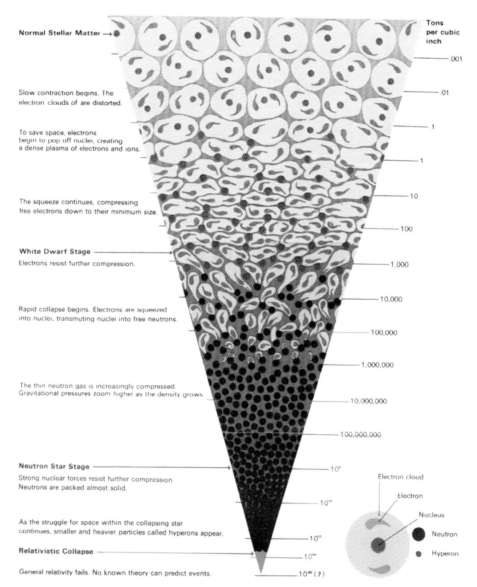

Normal Stellar Matter →

Slow contraction begins. The electron clouds of are distorted.

To save space, electrons begin to pop off nuclei, creating a dense plasma of electrons and ions.

The squeeze continues, compressing free electrons down to their minimum size.

White Dwarf Stage
Electrons resist further compression.

Rapid collapse begins. Electrons are squeezed into nuclei, transmuting nuclei into free neutrons.

The thin neutron gas is increasingly compressed. Gravitational pressures zoom higher as the density grows.

Neutron Star Stage
Strong nuclear forces resist further compression. Neutrons are packed almost solid.

As the struggle for space within the collapsing star continues, smaller and heavier particles called hyperons appear.

Relativistic Collapse

General relativity fails. No known theory can predict events.

Tons per cubic inch
.001
.01
.1
1
10
100
1,000
10,000
100,000
1,000,000
10,000,000
100,000,000
10^9
10^{10}
10^{11}
10^{12}
10^{∞} (?)

Electron cloud
Electron
Nucleus
Neutron
Hyperon

FIGURE 100   *Matter's struggles for space in the crush of collapse. The heavier a particle, the smaller the volume it is able to occupy. Atoms in a shrinking star are thus crushed, in turn, to electrons and nuclei, neutrons, and increasingly heavy hyperons. Whether each can counter gravity depends on the star's mass: less than 1.4 suns, free electrons stop the slow compression; less than 2 suns, neutrons will halt the collapse; more than 2 suns, no particle is able to resist gravity. (Source: Science Year, The World Book Science Annual. Copyright © 1968, Field Enterprises Educational Corporation.)*

specifically the energy carried away by neutrinos and over what time these neutrinos would be emitted. Among their predictions are those shown in Figure 101, which simply shows the calculated duration of the burning of the various elements in turn.

We were singularly fortunate that in 1987, when supernova 1987 A occurred in the Large Magellanic Cloud, one of our satellite minigalaxy detectors, originally constructed in the hope of observing the decay of protons and sensitive to neutrinos, was operational. Figure 102 shows the actual neutrinos detected in the Kamiokande detector outside Tokyo plotted as energy versus arrival time in seconds. Both the energy and the time characteristics confirmed the Brown–Bethe calculations and provided the first experimental proof of stellar collapse. As indicated, the supernova emitted something like 10^{57} neutrinos in a few seconds, ten times the total number of neutrons and protons in the sun.

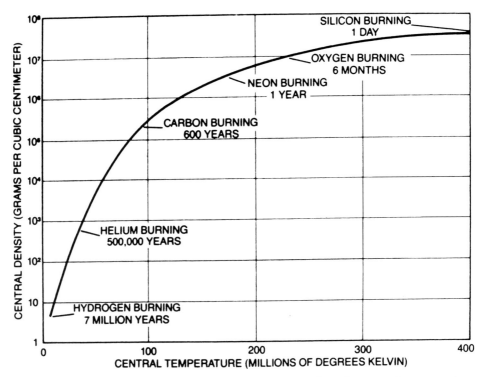

FIGURE 101   *The evolution of a massive star—one many times the mass of our sun—is a steadily accelerating progress toward higher core temperature and density. In this figure, the Brown–Bethe model of stellar development and explosion has been calculated for a star of 25 solar masses by T.A. Weaver of the Livermore National Laboratory. What is plotted is the central density against the central temperature and the various stages and durations of the dominant thermonuclear burning reactions are indicated along the curve. (Courtesy of G.E. Brown, SUNY, Stony Brook)*

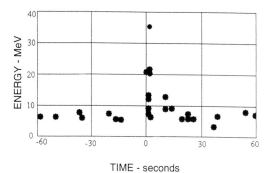

NEUTRINOS FROM SUPERNOVA 1987A
Data from Tokyo-Pennsylvania Collaboration
Kamiokande Detector

- Supernova emitted 10^{57} neutrinos in a few seconds -10 times the total number of neutrons and protons in the Sun

- Of these, about 10^{16} passed through the detector and 12 were detected

- Provides the first experimental proof of stellar collapse

- Pulse shape remarkably well reproduced by Bethe-Brown theoretical model based on nuclear phenomena

- Pulse duration sets upper limit of 15 eV on the neutrino mass.

FIGURE 102

Of these about 10^{16} passed through the Kamiokande detector and 12 were actually detected. Because of the work that had been done prior to this event, an absolutely amazing amount of information was obtained from these 12 detected neutrinos. Figure 103 is a photograph of the 1987 A supernova region as it exists today, and Figure 104 shows the interior of the new Kamiokande detector that became operational in 2000.

Gravitational Radiation

Theory predicts that supernovae should also emit large numbers of gravitons— or, put otherwise, generate large ripples in gravity that are radiated out from the collapsing star. These gravitons have not yet been detected, although the large laser interferometric detectors (LIGO), when operational, should be able to detect such events anywhere within our galactic neighborhood. As yet, the only concrete experimental determination of the reality of gravitational radiation has come indirectly, from the work of Taylor and Hulse of the University

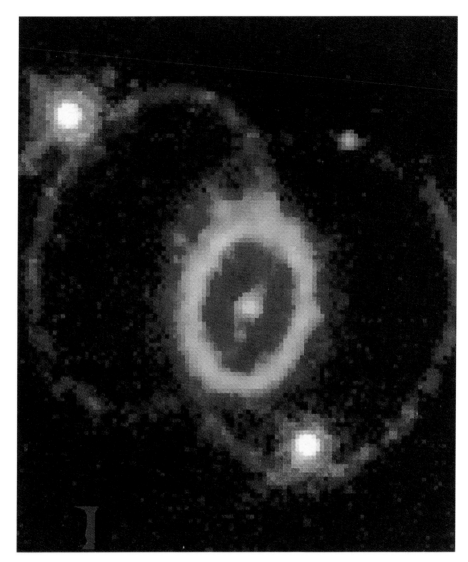

FIGURE 103   *The neutron star remaining after the supernova explosion of 1987A some 169,000 light years distant in the Large Magellanic Cloud illuminates rings of material thrown off from the star either during the supernova explosion or before. The bright ring is thought to be the outer part of the original star blown off during the supernova explosion, while the other two larger rings are not interlocking but in front of and behind the neutron star from our vantage point. 1987A was the first supernova for which detailed observations were possible throughout its original explosion and subsequent evolution. (Courtesy Robert Kirshner, Harvard University.)*

FIGURE 104   *The Super KamioKande detector during its filling. As it was filled with purified water, technicians used a raft on that water to systematically clean each of the 11,200 photomultipliers that will subsequently watch the 50 kilotons of water looking for Cerenkov light from neutrino scattering or proton decay. This facility, which sits in the Kamioka Zinc mine in the mountains west of Tokyo, is the world's largest Cerenkov detector. In a forthcoming experiment, this detector will be irradiated by neutrinos from the old 12 GeV KEK proton synchrotron in Tsukuba, 250 km away. This is of particular interest in view of the recent evidence of possible neutrino oscillation and finite mass obtained from comparison of the neutrino flux directly from the sun and after having passed through the earth on its way from the sun. (Courtesy of M. Koshiba, University of Tokyo.)*

of Massachusetts at Amherst, Massachusetts and now at Princeton University. In systematic observations of binary stars, using relatively crude radio telescopes, they were able to show that the rate at which the orbital phase shift was changing with time was precisely as predicted by the general relativistic calculations. Figure 105 shows a comparison between the predicted change in this parameter and the experimental measurements over a 12 year period. The agreement is remarkable. Gravitons will soon be detected.

The Evolution of Our Universe

Finally, let me turn to the grandest challenge of all. What can we, as physicists, say definitively about the evolution of our universe? Even a few decades ago, this was considered to be a question better suited to metaphysics or theological meditation rather than one properly proposed in physics. Happily, over the last decade and more, an amazing amount of information has become available and we can speak with considerable certainty about the origin and evolution of our universe.

Hubbell's measurements in 1917 refuted much prior theological speculation and showed that our universe was expanding, and from that expansion he deduced that its age was between 10 and 20 billion years. This, however, remained a rather isolated fact until in 1964 Arnold Penzias and Robert Wilson

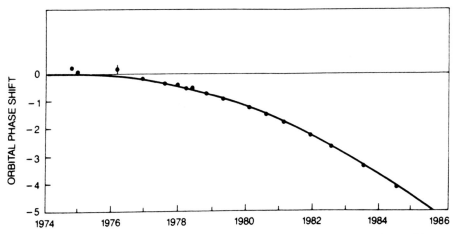

FIGURE 105   *Evidence for gravitational radiation from a binary pulsar as observed by Taylor and Hulse (a discovery which earned them the 1993 Nobel Prize in Physics). This figure shows the systematic shift in the pulsar's orbital period over the intervening years. The shift is almost precisely reproduced by calculations of the amount of energy and angular momentum that gravitational radiation would be expected to carry away from such a system. (Courtesy of A.D. Jeffries, P.R. Saulson, R.E. Spiro, and M.E. Zucker.)*

of the Bell Laboratories more or less by accident made one of the most important cosmological discoveries of the century—the fact that we are surrounded by an apparently uniform glow of microwave radiation. They were quite unaware of the importance of their discovery until a colloquium visitor happened to pass through Princeton and Bell Laboratories on consecutive days and, having spoken at length with Robert Dicke in Princeton about the fact there *should* be such a microwave background, he was able to tell Penzias and Wilson what they were actually observing. The detailed measurements of this cosmic microwave background (CMB) since its discovery have now provided very convincing evidence for the Big Bang origin of our universe and vital information on what happened after the instant of creation.

The primordial fireball at the instant of creation was a thick soup of elementary particles and light. As the fireball expanded and cooled, a majority of the particles annihilated one another, leaving light behind; and when the universe was 4 seconds old, the temperature had dropped to a point where even the lightest known stable particles, electrons and positrons, could no longer be created. It is this light from the primordial fireball that we see today as the CMB. For reasons that we do not yet understand, there was a tiny imbalance in the density of particles and antiparticles, so that a residual trace of matter—about one proton for every billion photons—remained. By the time the universe was three minutes old, it had cooled sufficiently to allow protons and neutrons to fuse into nuclei—deuterons, helium, and a little lithium and perhaps even a little boron. At the end of about 20 minutes, this primordial nucleogenesis had run its course, and nothing much happened for the next 300,000 years while the matter in the universe remained an opaque, ionized plasma of electrons and these light nuclei.

At about 300,000 years, the universe had cooled enough so that the electrons and nuclei could begin to form stable atoms, and this immediately made the universe transparent—thus decoupling the CMB from the matter. When we observe the CMB today, we are seeing photons created in the first minute of the universe that finally emerged from the primordial plasma when light and matter decoupled some 10–20 billion years ago.

The COBE Measurements on the CMB

There were early attempts to measure the spectrum of the CMB, but these are very difficult measurements and definitive results were not obtained until 1990 when NASA orbited its cosmic background explorer (COBE) satellite. Figure 106 shows the results obtained. The solid line is the predicted spectral distribution for a black body at a temperature of 2.728 K, and the fit to the experimental points is almost astonishingly good. Max Planck would have been pleased! This remarkable fit tells us two things: First, that the same laws of physics derived to explain phenomena observed in the laboratory today apply equally well to what was happening in the early universe; Second, that the CMB is a true fossil of the early universe apparently unchanged by everything that has happened except for the expansion-caused cooling. The COBE measurements of the CMB had a resolution of about 20 millionths of a degree, and small fluctuations of temperature in different directions were detected (at this level).

Measurements recently in the Python and Viper experiments in the Center for Astrophysical Research in Antarctica series have confirmed these tiny fluctuations in the CMB and have found them to be consistent with a flat universe, one having a so-called critical density of matter and energy—a key prediction of the best accepted theories of how the Big Bang got started. This is the inflationary universe proposed originally by Alan Guth of MIT and shown schematically in Figure 107.

While the Big Bang model accounts naturally for the expansion of the universe, for the existence of the CMB with its precise black body spectrum, and for the observed relative abundance of the light nuclei that emerged from the Big Bang, it does have remaining difficulties. It cannot explain why different parts of the universe do not have different CMB temperatures; it cannot explain why the universe today is anything but homogenous and it cannot ex-

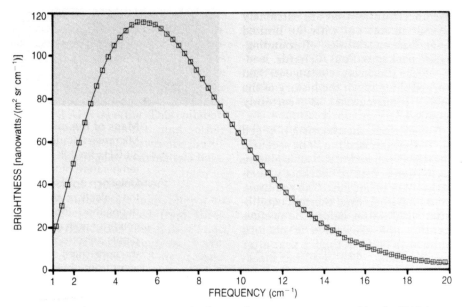

FIGURE 106   *The cosmic microwave background spectrum as measured by the NASA COBE satellite fitted with a black body spectrum for a temperature of 2,728 K. The size of the experimental point is far greater than the actual precision of the measurements, and they are enlarged here to improve their visibility. (Courtesy Richard Truly, NASA.)*

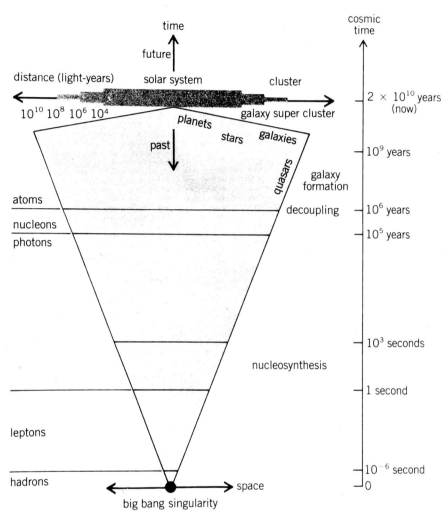

The inflationary universe

FIGURE 107 *A schematic illustration of the Guth inflationary universe with a variety of pertinent time and space parameters. In essence, this is a map of the behavior of our universe from the initial Big Bang to the present. (Courtesy of Alan Guth, MIT.)*

plain how the universe can be flat when it predicts a baryon density of about $\Omega=0.05$, while $\Omega=1$ for a flat universe.

The Inflationary Universe

In 1980, Guth proposed that the very early universe experienced a period of intense pressure, a kind of negative gravity, which resulted in an incredible expansion at a velocity far beyond our measured velocity of light. Few physicists have recognized that although the standard velocity of light as measured today is the limiting velocity within a fixed space, space-time itself has no such limit and can be expanded at transluminal velocities. In Guth's proposal, the expansion of the universe was extraordinarily brief, ending when the universe was about 10^{-35} seconds old; because of the intense pressure, however, the size of the universe had increased by a factor of roughly 10^{23}. Regions of the universe that had been in causal contact with one another were pushed apart with velocities substantially greater than that of light so that there could be no information transfer between them. Moreover, as the universe inflated, its density was driven to the exact value $\Omega = 1$.

At the end of the inflationary period, the universe was about the size of a grapefruit! But the tiny fluctuations that would eventually give rise to galaxies and groups of galaxies were now frozen in and out of communication with one another. Figure 108 taken by COBE shows the fine structure of the CMB predicted for the densities $\Omega = 1$ at the lower left and $\Omega = 0.1$ on the lower right. This provides the best evidence yet for a flat universe, one that will continue to expand indefinitely but with a constantly decreasing expansion of velocity. Recent data have suggested that, in fact, the rate of expansion is accelerating and this, if confirmed, might be a signature for a remnant of Guth's negative gravity. Indeed, very recently it has been suggested that this acceleration may provide evidence for a negative energy pervading the entire universe.

The Structure of the Milky Way Galaxy

Focusing for a moment on our own Milky Way Galaxy, we recognize that dust clouds make it impossible for us to see the central region of this galaxy in any visible wavelength; however, at radio wavelengths these dust clouds are transparent and we can see the central region, as shown in Figure 109. Detailed measurements of the velocities of stars in the vicinity of Sagittarius A, shown in larger scale in Figure 110, suggest that there is a black hole with some 50–100 million solar masses at the center of our galaxy. Indeed, measurements of other galaxies suggest that this is not only a common phenomenon but that perhaps our black hole is a relatively small one.

The Gross Structure of Our Universe

Moving away from our own galaxy to the visible universe more generally, in the last decade measurements made particularly by Gellar, Huchra, and Lapporent at Harvard have begun to show the gross structure of the universe. As shown in Figure 111, it is anything but homogeneous. We are located at the apex of this wedge, a wedge that would occupy about one degree in the map of the CMB, but

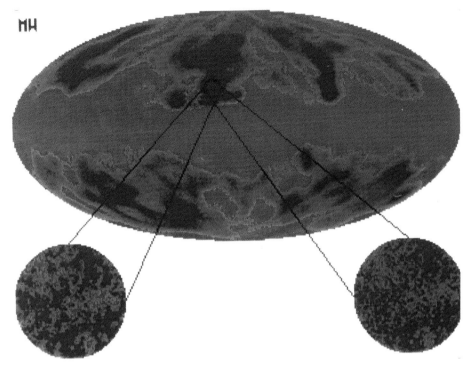

FIGURE 108   *The fine scale structure of the CMB predicted for universes with densities*
$\Omega = 1$ *(on the left) and* $\Omega = 0.1$ *(on the right). The two insets show the type of structure
that would be seen in a single 7-degree diameter patch of the sky (equivalent to the angular resolution of COBE) for each universe. The increased structure at smaller scale in
the* $\Omega = 1$ *universe is immediately evident. Mathematical analysis of a map of the entire
sky with this angular resolution and sensitivity should provide enough information to determine the total density of the universe (and thus determine whether it will expand for-
ever) and what fraction, if any, of the total density of the universe is in the form of dark
matter. (Courtesy Andrew Lange and Martin Hoyle, California Institute of Technology.)*

already we see large clusters of galaxies and large open voids. These are remnants
of the tiny fluctuations in the density of the original universe prior to inflation.

Out of all of this comes a major and remarkable puzzle. If our universe is
indeed flat, then we see less than 10% of the total mass in any of our measurements, and more than 90% of the mass of the entire universe is in what is
now called dark matter. Some of this dark matter could clearly be composed of
brown dwarfs, protostars that were too small to ever ignite and that are therefore not visible. If the neutrino has a nonzero mass, it too can contribute to the
dark matter and the possibility always exists that there are exotic forms of matter which we have never knowingly encountered. Of one thing we are confident, however; all this dark matter should be observable through its gravita-

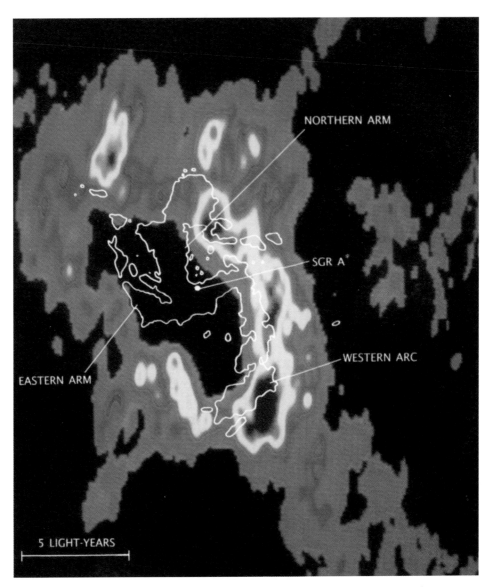

FIGURE 109 *Hydrogen cyanide molecules in warm dense gas clouds near the galactic center emit short wavelength radio emissions, shown in this map made with the University of California's millimeter wavelength telescope. Colder gas farther out is not visible at these wavelengths. An elliptical distribution of gas with a hollowed-out center lies at the heart of the galaxy. The gas is clumpy and is varied in its intensity of radiation, indicating that it has been disturbed recently, most likely within the past 100,000 years. Collisions between clumps will eventually smooth out the gas. (Courtesy of the University of California at Berkeley.)*

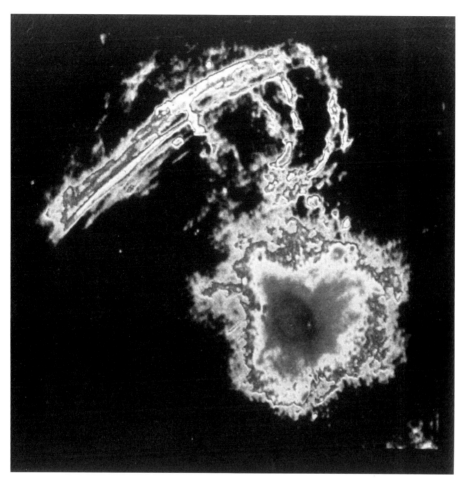

FIGURE 110   *Shown here at 20cm wavelength is a radio map of Sagittarius A, at the lower right, and an extensive network of filaments looping toward the upper part of the figure. The galactic center is the bright point in the center of the Sagittarius A region. None of this structure, and indeed nothing in the region of the galactic center, is open to visible light observation because dense intervening dust clouds in the plane of the galaxy block out our line of sight. Very recent data have suggested that the total mass of the universe is 5% visible matter, 35% dark matter, and 60% dark energy! (Courtesy of the National Radio Astronomy Observatory.)*

tional effects, and this has led to major interest in the phenomenon know as gravitational lensing.

Gravitational Lensing

Concentration of matter on our line of sight to quasars 10–20 billion light years distant can act as lenses, and from the observation of the lens effects, we can determine the masses involved in the lensing. Figure 112 is a typical gravita-

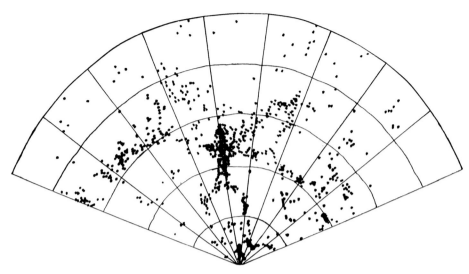

FIGURE 111   *The distribution of galaxies extending from us outward to a radius of approximately 600 million light years. A region of space this big would subtend about one degree on the map of the CMB. We are located at the apex of the wedge at the bottom center of the figure. The map is constructed by using the radial velocity of each galaxy to estimate its distance. The torso and head of the rather strange human-like figure in the center of the map is a large cluster of galaxies. Its apparent elongation along the radial direction is an artifact of the map-making technique and reflects the orbital velocities of the galaxies withing the cluster. The distribution of galaxies remains highly inhomogeneous on scales up to hundreds of millions of light years. The contrast between the near isotropy of the CMB and the structure observed in this map poses a puzzle that the inflationary model resolves: how did these structures form from the nearly homogenous early universe? It bears noting that, at least from this vantage point, the galactic distribution has the general appearance of large bubbles of essentially empty space confined by walls comprising galactic clusters. The long structure that extends from the "arms" of the central figure is the largest structure yet identified in our universe. This map has been constructed from data originally taken in 1986 by Lapporent, Gellar, and Huzhra. (Courtesy of Andrew Lange and Margaret Gellar, California Tech and Harvard.)*

tional lens effect showing four separate images of a distant quasar symmetrically located around a mass concentration. There are a very large number of such situations, and instrumentation is now being developed to scan for them rapidly and efficiently in the hope of coming up with a better estimate of the amount of dark matter out there.

There is a related phenomenon known as microlensing, which may allow us to observe small concentrations of dark matter in the vicinity of our own galaxy. Observation of what may well be microlensing is shown in Figure 113; the idea is simply that if we were to maintain a constant vigil on the night sky, we might discover that from time to time, as a concentrated dark matter entity crossed our line of sight to an individual star, gravitational lensing could make it appear to shine much more brightly than normal during the time of passage through the line of sight. This may be the case illustrated in Figure 113, and several groups are considering the establishment of long-term monitoring of the night sky to search for more examples.

Concluding Remarks

I conclude with a list, in Figure 114, of some 10 open questions in physics. There are, of course, many more such questions, but these are ones that strike me as being of particular interest and I leave them to the physics of tomorrow.

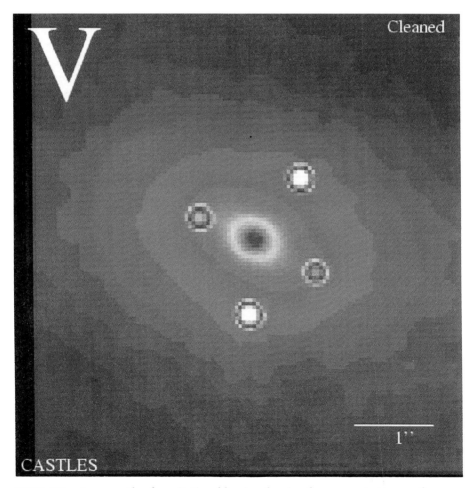

FIGURE 112   *An example of gravitational lensing showing four separate images of a quasar whose line of sight (to us) is interrupted by a large mass concentration. (Courtesy of Charles Baltay, Yale University.)*

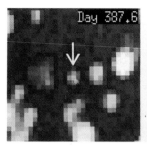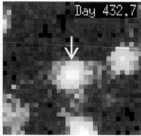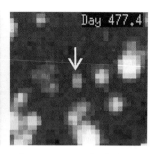

FIGURE 113 *A possible example of microlensing in which an object composed of dark matter may have crossed the line of sight between observers on the earth and the star in question. The symmetrical brightening and dimming of the image as the object passes through the line of sight are characteristic of the microlensing effect. Currently, three groups of investigators are searching for these events. They include the U.S. Australian team led by Charles Alcot of the Lawrence Livermore National Laboratory, the U.S-Polish group led by Bodhan Paczynski of Princeton University, and a French collaboration headed by Michele Spiro of the Saclay Research Center. Alcot's group, beginning in 1993, has been monitoring 1.8 million stars in the Large Magellanic Cloud and for nearly a year saw nothing whatever before the event illustrated here showed up. It was recorded simultaneously by the French group and results were published jointly. Recently, Paczynski has reported yet a third example involving a star towards the center of the galaxy. It is far from certain that the observed events thus far really do result from gravitational interaction with dark matter, but the fact remains that dark matter—by its very nature—can only be "seen" through gravitational interaction. Fortunately, the instrumentation is now available that makes it possible to observe very large numbers of stars on a continuing basis. As indicated in these figures, that on the left was taken on day 387.6, that in the middle on day 432.7, and that on the right on day 477.4 of the experimental observations. The instrumentation uses digital detection, hence the discrete pixels appearing in this figure. (Courtesy of Charles Alcot, Lawrence Livermore Laboratory.)*

More than a century ago, Lord Raleigh, then president of the British Association for the Advancement of Science, was asked to give a review of physics in the 19th century as his presidential address to the Association meeting in Montreal. He began by noting that this was impossible, and I know only too well how he felt. But I can do little better than to quote one of his closing comments, which appears in Figure 115. "Increasing knowledge brings with it increasing power, and great as are the triumphs of the present century we may well believe that they are but a foretaste of what discovery and invention have yet in store for mankind." To "discovery and invention" I would perhaps have added "surprises." We have had many of them in the twentieth Century and I very much hope that there will be many of them waiting for us in the years ahead. and Figure 117 is unfortunately reminiscent of a situation with which all of us have become familiar and

SOME OPEN QUESTIONS IN PHYSICS

1. **HOW DOES MASS ORIGINATE? HIGGS FIELD?**

2. **DOES NONBARYONIC DARK MATTER EXIST? WHAT FORM?**

3. **WHY ARE WE IN A MATTER UNIVERSE?**

4. **WHAT IS THE ULTIMATE FATE OF OUR UNIVERSE?**

5. **WHAT IS THE STRUCTURE OF QUANTUM GRAVITY?**

6. **ARE QUARKS AND LEPTONS TRULY ELEMENTARY — OR COMPOSITE?**

7. **DO THE PHYSICAL CONSTANTS CHANGE WITH TIME?**

8. **WHAT ARE THE CONSEQUENCES OF A NONZERO NEUTRINO MASS?**

9. **HOW DOES ONE BUILD A QUANTUM COMPUTER?**

10. **IS ROOM TEMPERATURE SUPERCONDUCTIVITY POSSIBLE?**

FIGURE 114

with which we will continue to live as the competition for investment in science and technology becomes more intense and the need for coherence in the arguments we make to our governments for continued investment becomes ever greater.

We remain a vital, active, and productive science. We physicists are among the most fortunate of humans: we have been privileged to engage in that greatest adventure of discovery at a time when technology has allowed us to push

"Increasing knowledge brings with it increasing power and great as are the triumphs of the present century we may well believe that they are but a foretaste of what discovery and invention have yet in store for mankind."

Lord Raleigh, 1884
A Review of Physics in the 19th Century
Presidential Address to the British Association

FIGURE 115

"*Three ships is a lot of ships. Why can't you prove the world is round with <u>one</u> ship?*"

FIGURE 116   *(Copyright The New Yorker Collection 1970 Frank Modell from cartoonbank.com. All Rights Reserved.)*

outward the frontiers of knowledge at unprecedented rates. And in so doing we have also bettered the lives of humans everywhere. Physics, as the most fundamental of the sciences, will always remain a vital part of this great adventure.